Paddle
Warships

Virago.

Virago was typical of later paddle vessels. She was a first class sloop and was derived from John Edye's very successful *Gorgon* design, though she had bigger engines.
(D K Brown collection)

Paddle Warships

The Earliest Steam Powered
Fighting Ships
1815-1850

David K Brown RCNC

Illustrated with draughts from the collections of
the National Maritime Museum, Greenwich

CONWAY
MARITIME PRESS

First published in Great Britain in 1993 by
Conway Maritime Press, an imprint of Brassey's (UK) Ltd,
101 Fleet Street,
London EC4Y 1DE

British Library Cataloguing in Publication Data
  Brown, D. K.
    Paddle Warships: Earliest Naval Steamers
    1815-51. — (Ship Types Series)
    I. Title   II. Series
    623.8

  ISBN 0 85177 616 7

Designed and typeset by Tony Garrett, Fathom Graphics, Exeter

Printed and bound in Great Britain by The Bath Press, Bath

Contents

Introduction

THIS BOOK FOLLOWS the pattern set by Robert Gardiner in *The First Frigates* as part of a series intended to provide design histories for a range of ship types which have not previously been fully described in published works. It follows the style of the earlier book with a chronological record of the development of design philosophy in Part I including tables of ship specifications, building data, machinery and armament particulars.

The mainstream of paddle warship progress is generally clear, with steady evolution, though it is a bit confused by a number of ships built for ancillary duties, including surveying and exploration in the colonies and, in the later years, by a number of experimental designs. Even the Admiralty found it difficult to classify the increasing number of paddle warships of widely differing fighting capability. Part I concludes with a tabulation of some of the classifications adopted, in most cases associated with a desire to standardise armament, rig and complements.

Part II considers the evolution of specific aspects of design such as layout, structure, rig etc, again with a number of tables presenting information which is not readily available elsewhere. As with Gardiner's book, the illustrations are mainly selected from the Ships Plans collection at the National Maritime Museum, Greenwich, and a complete list of plans available there is included at the end of the book.

The paddle warship

The Admiralty began building paddle steamships in the early 1820s, at first as tugs to tow sailing warships out of harbour in a calm or against a contrary wind. It was also envisaged that they could tow sailing ships into action in unfavourable winds, a role finally fulfilled in the bombardment of Sevastopol in 1854. The free mobility of the steamship, independent of wind, led to their use in coastal operations for which they were given a light armament.

From about 1830, bigger ships were built, with a powerful armament, a trend which continued until the late 1840s, by which time the big paddle frigates were fast and very heavily armed. They fought no battles of significance but the Royal Navy and others were able to appreciate the importance of the steamship which, tactically, could move freely in a calm or against the wind and, strategically, could be relied on to arrive at its destination at a planned time.

These early steamships were not without problems; in particular they depended on frequent supplies of coal, not always easy to obtain in distant waters, and they were unable to make long ocean passages under steam alone. Though the machinery was generally reliable, it could and did break down but, with such simple engines, repairs were usually within the capability of a ship's crew. The big paddle wheels and their sponsons very considerably reduced the number of guns which could be carried on the broadside whilst the heavy machinery and coal took up about a third of the internal space. The machinery cost as much as the ship, which made their total cost about double that of a sailing ship of similar size.

Though there are numerous legends of opposition to steamers, the Admiralty files contain many letters from Commanders-in-Chief asking for more steamships, whilst a growing number of young officers sought to improve their knowledge of the steam engine. Looking back, 150 years later, there is a tendency to think of these ships as quaint, an evolutionary blind alley but, to contemporaries, they represented the advanced technology of the day.

Acknowledgements

When I first became interested in paddle warships some twenty years ago – as a relief from my duties as a propeller designer – I found that they were hardly mentioned in any book and the few authors who did refer to them saw them as of little value and a blind alley. My enquiries soon led me to the late George Osbon of the National Maritime Museum, who provided much of the tabular data in this book. In particular, the annex on Post Office packets is derived almost entirely from his research. The Osbon papers are now owned by the World Ship Society and I am most grateful to their naval secretary, Richard Osborne, for allowing me to refer to, and to quote from, these notes.

Thanks are also due to others at the National Maritime Museum: David Lyon, as always a wonderful source of information; David Topliss and his colleague, Graham Slatter, who helped so much in selecting the plans. Joe Room, formerly of the Science Museum, has been most helpful in supplying material on early steam engines and I am grateful to the Trustees of the Museum for permission to use some illustrations. Finally, Robert Gardiner, for suggesting the subject and for making drafts of his book available to assist in planning this one. It is a short list of helpers: even today, few are knowledgeable concerning paddle warships.

Part I: Design History

Lightning.
Many paddle warships had very long active lives. This model shows *Lightning*,
built in 1823, as the survey ship for the Baltic Fleet in 1854.

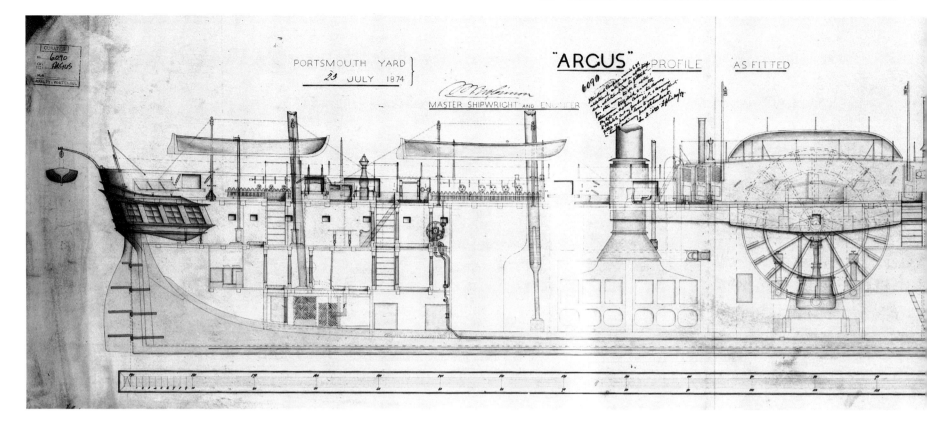

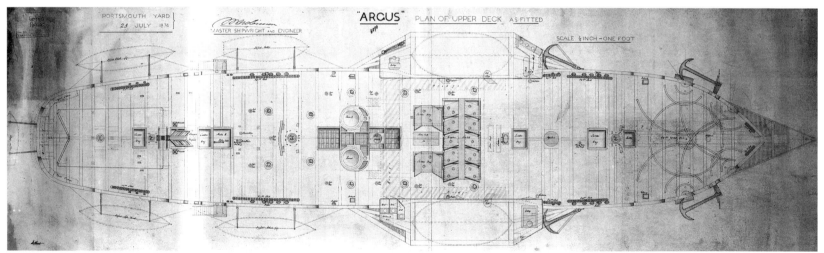

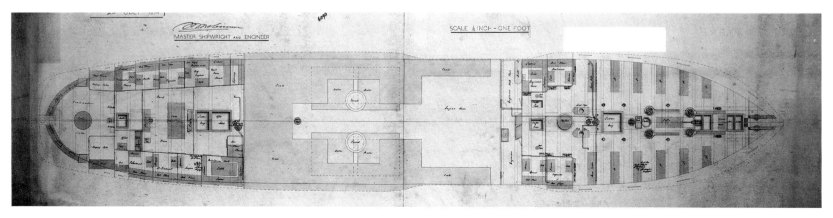

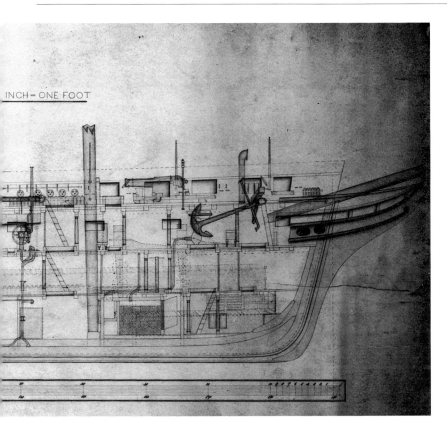

INCH – ONE FOOT

Argus was one of the last paddle sloops, designed by Fincham in 1847. However the layout changed little over the years, since it was dominated by the space required for the engines and boilers amidships, and this arrangement is typical of most paddle warships.

Some points to note are: the length required by the machinery and the obstruction of the side by paddle wheels and sponsons; the galley and its chimney at the fore end of the starboard sponson. The large and very beamy boats over the box. *Argus* normally carried a large pivot gun at each end and four carriage guns amidships, but this drawing of 1874 appears to show only one gun forward. The main mast is stepped in the boiler room on an iron base.

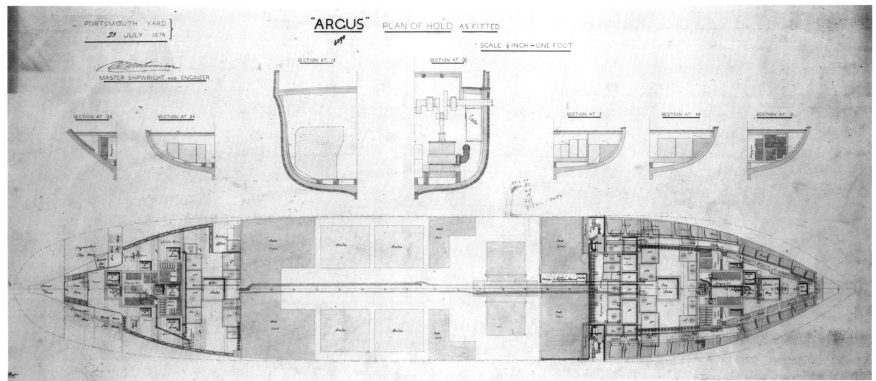

1. Background: Early Steam Warships

THE FIRST BRITISH attempt at a steam warship was the *Kent* of 1793, conceived by the Earl of Stanhope[1] and built at his own expense. It was intended to propel her by six feathering paddles on either side (not wheels), a scheme which proved to work when they were operated by hand. Her engine, by James Watt, was apparently a failure and she did not enter service as a steamship.

Robert Fulton was responsible for many of the early US commercial steamships and in 1814 Congress approved his plan for a warship driven by a single paddle wheel on the centreline.[2] *Demologos* proved very slow

Lightning, the Royal Navy's second steamship, ordered in 1823.

There are no plans of *Comet*, the first Royal Navy steam warship, of 1821. *Lightning* had a long and active life; deploying to Algiers in 1824, to the Baltic in 1854 and only broken up in 1872. Oliver Lang was responsible for almost all the early steamships and deserves high praise. Both a profile and an upper deck plan of *Lightning* are shown in this page.

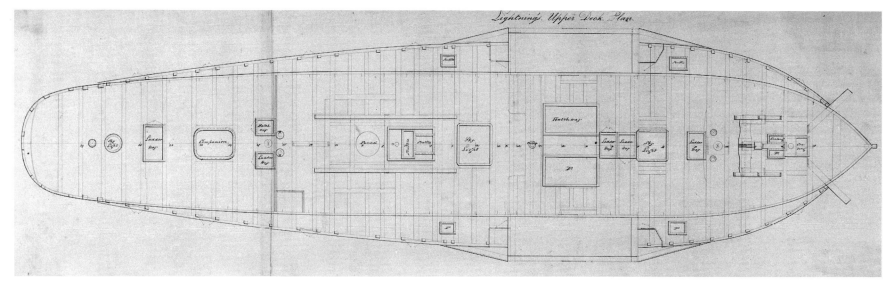

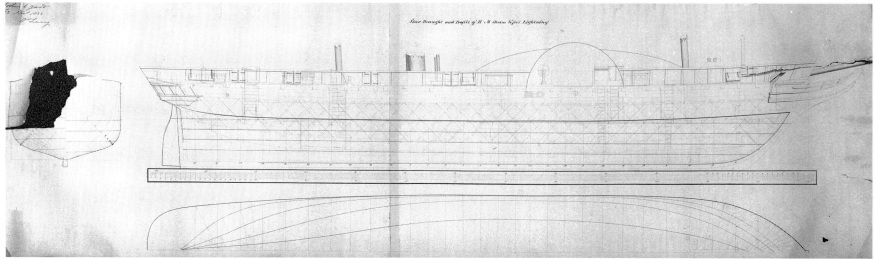

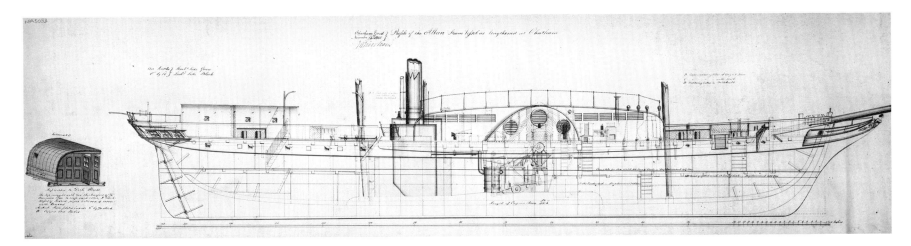

on trials in 1815 and with the end of the war and the death of Fulton interest waned and she did not go to sea again.

By the end of the war in 1815, commercial steamships were appearing in considerable numbers in British waters and when, in that year, an expedition was planned to the Congo river, the Board was easily persuaded by James Rennie, then working for the Admiralty as the engineer for Plymouth breakwater, to build a steamship, the *Congo*. She completed in 1816 but was said to be so overweight that her engines had to be removed.

Despite this setback, the Board retained an interest in steam ships, inspired by the Second Secretary (senior civil servant) John Barrow, a supporter both of exploration and of technical innovation. Marc Brunel, who had done much to bring the industrial revolution into the Royal Dockyards, persuaded the Board to charter commercial steamships in 1816 and 1819 for trials involving towing sailing warships out of harbour against the wind. The trials seem to have been only partially successful as the power of contemporary steamers was marginal for such work.

The Post Office, then responsible for the mail packet service, (Appendix) ordered two ships from Elias Evans of Rotherhithe, to be built under Admiralty supervision. These two ships, *Lightning* and *Meteor*, entered service on the Holyhead route in 1821 and proved very successful. Later that year, *Lightning* carried King George IV to Ireland when sailing ships were becalmed, so publicising the steamship. After many changes of name, she entered Admiralty service as the *Monkey* in 1837 and is often incorrectly described as the first Admiralty steamship.

Early Admiralty Steamships

In the same year, 1821, the Admiralty ordered its first steam vessel, the *Comet*, to work on the Thames as a tug and ferry. The Surveyor, Sir Robert Seppings, delegated the design to Oliver Lang, Master Shipwright at Woolwich, who adapted the lines of a sailing sloop for her. The plans have not survived but she is known to have had a single tall funnel, circular paddle boxes without sponsons, a knee bow, square stern and to have been rigged as a two-masted schooner.

Comet had a side lever engine of 80 Nominal Horse Power (NHP)[3]

Alban shown after lengthening in 1831.

She was one of a class of five ordered in 1824, showing the Admiralty's early recognition of the value of steam ships. Note the elaborate deckhouse added aft.

Table 1: EARLY STEAM WARSHIPS
Specification

Name	Tons bm	Disp't	Length ft in	Beam ft in	Depth ft in	Draught For'd ft in	Aft ft in
Congo	86		70 0	16 0	8 10	4 3 (mean)	
Comet	238	239	115 0	21 3	11 11	7 0	7 6 (See text)
Lightning and Meteor	296	349	126 0	22 8	13 8	9 0	9 6
Alban and Carron	295		109 8	24 10	13 6	11 2	12 4
African	295		110 0	24 11	13 4	11 2	12 4
Echo and Confiance	295		111 9	24 10	13 6	11 2	12 4
Alban (Lengthened)		407	145 0	24 10	13 6	11 2	12 4
Columbia	361	504	130 6	24 9	13 7	10 5	11 7
Hermes (ex-Courier) and Messenger	733	912	155 6	32 9	12 0	10 3	10 9
Firebrand and Flamer	495	510	155 3	26 5	14 10	9 9 (mean)	
Black Eagle	540	715	168 0	26 5	14 10	9 9 (mean)	
Pluto	365	386	135 0	24 3	11 10	6 6	6 6
Dee	704	907	166 7	30 4½ 51 0 oa	16 4	11 6	11 6
Firefly and Spitfire	550	677	155 0	27 8½	16 2	9 10	10 10
Blazer and Tartarus	523	560	145 0	28 1	14 9 15 1	10 6 (mean)	

Table 2: EARLY STEAM WARSHIPS
Building Data

Name	Builder	Laid Down	Launched	Completed	Fate
Congo	Deptford Dyd	Oct 1815	11 Jan 1816	16 Feb 1816	BU 1826
Comet	Deptford Dyd	Nov 1821	23 May 1822	13 July 1822	BU Oct 1868
Lightning	Deptford Dyd	Feb 1823	19 Sep 1823	1823	BU Jan 1872
Meteor	Deptford Dyd	Feb 1823	17 Feb 1824	1824	BU Aug 1849
Alban	Deptford Dyd	Jul 1824	27 Dec 1826	1831	BU May 1860
Carron	Deptford Dyd	Jul 1824	9 Jan 1827	1828	BU 1877
African	Woolwich Dyd	Sep 1824	30 Aug 1825	1825	BU Nov 1862
Echo	Woolwich Dyd	Dec 1824	28 May 1827	1827	BU Jan 1885
Confiance	Woolwich Dyd	Feb 1825	28 Mar 1827	1827	BU Jun 1873
Columbia	Woolwich Dyd	Mar 1827	1 Jul 1829	1829	Sold 29 Oct 1859
Courier (*Hermes* 24 Jun 1830)	Butterly (Blackwall)		1824	Both purchased 20 Aug 1830	BU Jun 1854
Messenger	Butterly		1824		BU 1861
Firebrand	Curling Limehouse	April 1831	11 Jul 1831	1831	BU Mar 1876
Flamer	Fletcher Limehouse	April 1831	11 Aug 1831	1831	22 Nov 1850, wrecked off W Africa
Pluto	Woolwich Dyd	Feb 1831	28 Apr 1831	1831	BU Mar 1861
Dee	Woolwich Dyd	1825 (as sail)	5 Apr 1832	1832	BU 1871
Firefly	Woolwich Dyd	May 1832	29 Sep 1832	1834	BU 1866
Spitfire	Woolwich Dyd	Dec 1832	26 Mar 1834	1834	10 Sep 1842, wrecked Jamaica
Blazer	Chatham Dyd	Nov 1833	May 1834	1835	BU Aug 1853
Tartarus	Pembroke Dyd	Sep 1833	23 Jun 1834	1834	BU Nov 1860

driving 'common' paddle wheels 14ft in diameter which gave her a speed of about 7½ knots.[4] *Comet* spent her early years on the Thames and, with other early steamers, appeared in the Navy List for the first time in 1828 as 'HM Steam Vessel'. She carried a light armament of two, sometimes four, guns, usually 6-pounders. *Comet* cost £4,314 for the hull, and her machinery was a further £5,050. In 1839 it was reported that she '...labours in a heavy sea in consequence of her being deeper in the water than was intended at first by her constructor'.[5] While records are not clear, it looks as though her design draught (mean) was 7ft 3in but that she floated in service at nearly 9ft.

In 1837 she became a survey vessel but by 1854 was a tug again, in Portsmouth, where she remained until she was broken up in 1869. Lang seems to have realised the need for a strong hull to take both the concentrated weight of the machinery and to absorb the vibration of these early engines, since all his steamers had very long lives. *Comet* was clearly seen as a success, judged both by her long service life and by the number of vessels which were built based on her design.

The next two steam vessels, *Lightning* and *Meteor*, were ordered in 1823 and, though generally similar to *Comet*, were appreciably bigger. They, too, were intended as auxiliaries but, in 1824, *Lightning* joined the squadron for the expedition to Algiers, the first operational deployment of a steamship by the Royal Navy or any other navy. She appeared in the Navy List for the first time in January 1828. Much later in her career, *Lightning* achieved further fame as the survey vessel of

the Baltic Fleet in 1854[6] and she was only broken up in 1872. *Meteor* had a shorter life of twenty-five years.

Five more steamers of the *Alban* class were ordered in 1824, designed by Lang and of similar style to his earlier ships, though a little beamier and shorter. This comparatively large order goes far to disprove the oft quoted, but incorrect, story that the First Lord, Melville, was totally opposed to the steamship.[7]

Most of these early steam vessels were completed with a schooner rig but *Alban*, at least, was re-rigged as a brigantine. An early report said that she was a good seaboat but of insufficient power as built.[8] In 1831, she was lengthened by 36ft and given a new engine. At the same time, she was re-armed with one long pivot and two shorter 32-pounders, the first time a steam vessel had been given a fighting armament. Her sisters had their original armament of two 6-pounders increased to six 18-pounders. This class also had a very long life – the first being broken up in 1860 – though some spent their later years as hulks.

Funds for new construction were very limited and, in 1826, the Board decided to buy two existing ships for use as mail packets on the Corfu route. Money for the purchase only became available in 1839 and these two ships, *Courier* – almost immediately re-named *Hermes* – and *Messenger*, proved a bad buy, being discarded after only a short service life. The steamship service to Corfu was opened in 1830 by *Meteor* and was very successful, steamships being able regularly to complete the

Alban, as lengthened in 1831.
She was typical of Lang's early ships and, after she was lengthened, was perhaps the first paddle ship to carry a fighting armament.

round trip in a month whereas sailing packets took about three months.

In 1827 the Admiralty ordered the *Columbia*, larger but not very different from the earlier *Alban*s. She was completed with common paddle wheels but was later tried, first with Morgan's feathering wheels and then with cycloidal wheels. The *Dee* was also ordered in 1827, a much larger vessel with a displacement of 907 tons. She was modified by Lang from a sailing vessel designed by Seppings and was of much the same style as Lang's smaller ships though she was completed as a brigantine. Apparently, she was a poor seaboat, a report reading '...very slow and steers very wild being trimmed by the stern. Using the after coals first improves her. She rolls very deep and always uneasy. Cannot carry quarter boats in a moderate beam wind.'[9]

Dee became a troopship in 1855 with a topgallant forecastle for her crew and a poop and deckhouse aft. Later, in 1866, she became a store ship.

There was a change of government in November 1830 and Sir James Graham replaced Lord Melville as First Lord. The first two steamships ordered by the new Board were of Lang's familiar design but were built in commercial yards on the Thames. These ships are of interest due to

Table 3: EARLY STEAM WARSHIPS
Machinery Particulars

Ship	Engines by	NHP	Type	No Cyl	ihp = speed (trials)
Congo	Watt	20	Beam	2	5
Comet	Watt	80	Side lever	2	7.6
Lightning	Maudslay	100	Side lever	2	
Meteor	Watt	100			
Alban	Watt				
Carron	Watt				
African	Watt	100	Side lever	2	7.6
Echo	Watt				
Confiance	Maudslay				
Columbia	Maudslay	140	Side lever	2	8.0
Later	Watt	100	Side lever	2	
Hermes	Butterly	140	Side lever	2	
Messenger	Butterly	200			
Firebrand	Butterly	140	Side lever	2	(1832 Maudslay)
Flamer	Morgan	140	Side lever	2	8.5
Black Eagle	Penn	–	Oscillating	2	(1843)
Pluto	Watt	100	Side lever	2	7.5
Dee	Maudslay	200	Side lever	2	272 8.0
Firefly	Butterly		Side lever		(1843 Maudslay, 220 NHP, 2-cyl, 9.5 kts)
Spitfire	Butterly		Side lever		
Blazer	Miller	136	Side lever	2	
Tartarus	Maudslay	100	Side lever	2	

Sail plan of *Alban* after a mizzen was added.

It was hard to provide a balanced sail plan as the engines prevented the main mast being stepped in an optimum position. (See Chapter 11)

Dee, modified by Lang from a sailing vessel.

She was regarded as a poor seaboat and converted to a troopship in 1855 with a topgallant forecastle for the crew and a deckhouse aft, seen in this profile.

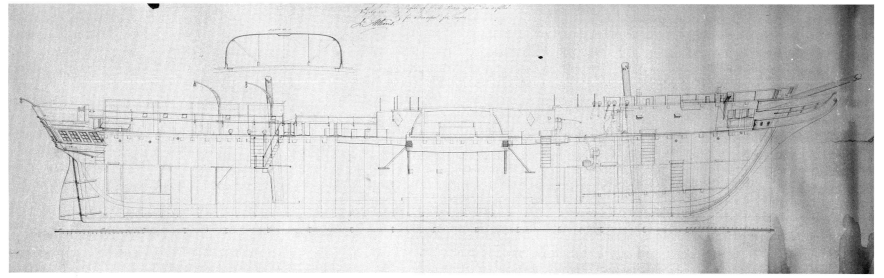

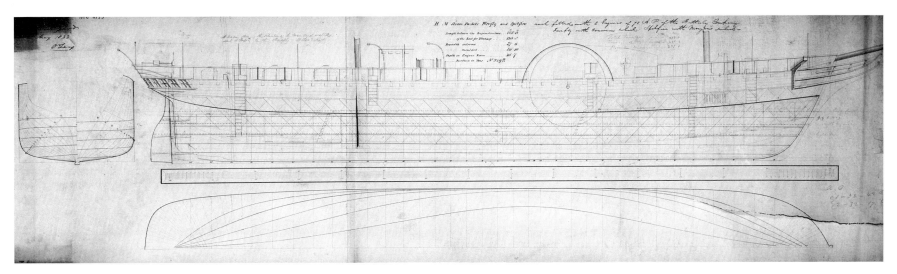

Firefly, typical of Lang's early paddle vessels.

Points to note are the small rise of floor in the sections, circular paddle boxes and the numerous diagonal iron straps strengthening the hull.

Black Eagle. Originally named *Firebrand*, she was renamed as the Admiralty yacht in 1843.

At the same time she was lengthened and given new engines and boilers. The boilers were arranged fore and aft of the engine room so that she now needed two funnels.

their frequent use for experimenting with new machinery. *Flamer* had one engine change in 1832 but retained her original feathering wheels. She was said to sail well 'particularly on a wind'.[10]

Firebrand was given a change of engines (ex-*Columbia*) in 1832. Engine changes were fairly common at this time, ships whose engines needed a lengthy overhaul being given reconditioned units from another ship. In 1835 *Firebrand* was used by the Board as a yacht to visit the Dockyards and in August 1843 she became the official Board yacht with a new name, *Black Eagle*, in honour of the Prussian Royal family who often borrowed her. More significantly, she was lengthened and given the first set of John Penn's oscillating cylinder engines which had the same weight and volume as the original set but were of twice the nominal power, though it does not seem that the power delivered to the paddles increased very much.

Black Eagle was given tubular boilers with 2250 brass tubes taking the hot gases from the furnace through the water space. The boilers were arranged fore and aft of the engines so that she had two widely

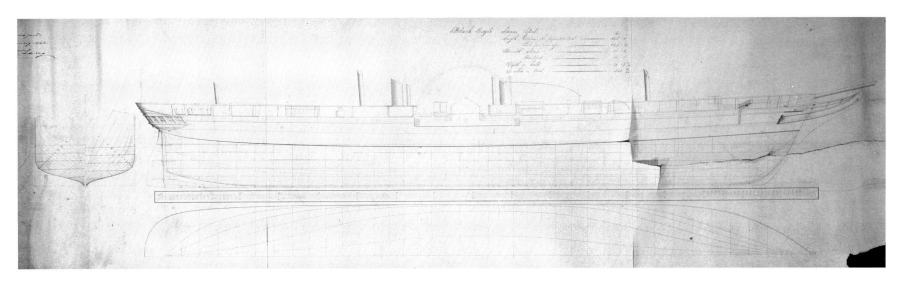

spaced funnels, probably the first two-funnelled warship. Much later, in 1856, she was fitted with Wethered's superheated steam boiler for trials.

Pluto was also ordered in 1831, to Lang's design, as a shallow draught (6ft 6in) vessel to fight piracy in the West Indies. Since most of the operations of these early steamers were in shallow water, it is surprising that they were not all given shallow draught even though it would have affected their sailing. A little later, in 1832-33, four more vessels of Lang's design were ordered. One of these, *Blazer*, was said to sail at 6.4 knots by the wind and 7.6 knots before the wind.

Perhaps rather arbitrarily, one may see 1831 as marking the end of the first period of steamship building. The operations off the Dutch coast in that year showed the value of the steamer in coastal blockade and, though many of the earlier ships were later given a heavy armament, from now on ships would complete with a numerous and powerful fit of guns. By 1831 the Royal Navy had fourteen steamships, of which eight were in commission, and there were frequent demands from sea for more.

Table 4: EARLY STEAM WARSHIPS
Armament

Where the date of a change in armament is not known, the outfits will be numbered 1, 2 etc. In these early ships, all guns were carried on the upper deck.

Ship	Date	Armament
Congo	Design	1x12pdr, not fitted
Comet	1	2xbrass 6pdr
	2	2x18pdr/15cwt
	3	3x9pdr/13.5cwt
Lightning and *Meteor*	}	2 (later 3) x 6pdr/6cwt brass
Alban *Carron* *African, Echo* and *Confiance*	1 (all 5) / 2 (all 5)	2x6pdr/6cwt brass / 4x18pdr/10cwt, 2x18pdr/10cwt carronades
Alban (lengthened)		1x32pdr/56cwt pivot, 2x32pdr/25cwt
Columbia	1	1x32pdr/25cwt pivot, 2x32pdr/25cwt
	2	1x32pdr/42cwt pivot, 2x32pdr/25cwt
Firebrand and *Flamer*	1 / 2	6x9pdr/13.5cwt / 1x32pdr/25cwt pivot, 2x32pdr/17cwt
Black Eagle		1x18pdr/22cwt pivot
Pluto	1	2x18pdr/22cwt
	2	1x32pdr/42cwt pivot, 2x32pdr/17cwt
Dee	1	2x18pdr/22cwt
	2	4x32pdr/63cwt pivot, 2x32pdr/56cwt pivot
	3	1x10in/86cwt pivot, 4x32 pdr/63cwt
Firefly	1	2x9pdr/13.5cwt
	2	1x32pdr/50cwt pivot, 2x32pdr/17cwt
Spitfire		1x32pdr/50cwt pivot, 4x32pdr/25cwt
Blazer and *Tartarus*	1 / 2	2x9pdr/13.5cwt / 1x32pdr/25cwt pivot, 2x32pdr/17cwt
	1856	1x32pdr/25cwt pivot, 2x32pdr/25cwt

Tartarus. **This profile shows the elaborate engine frames fitted in many of the early steam ships.**

The drawing also shows the heavy, close framing adopted by Lang in the machinery spaces to accept the concentrated weight and the vibration of these early engines.

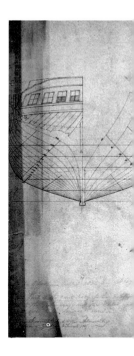

2. Transitional Ships

ALL THE EARLIER ships had been designed by Oliver Lang and, though they were successful, the Graham Board decided in 1831 to build four competitive designs, one to the Surveyor's design (Seppings) and the others by different Master Shipwrights. All four were successful and were very little different from the earlier ships. It is a tribute to Lang's skill and experience that his ship, *Medea*, formed the basis for so many later ships. Initially they had a schooner rig but later changed to barque or barquentine. Developments in machinery meant that when *Medea* was re-engined in 1846, the engine room could be reduced in length by 12ft even though the new unit was of the same NHP and weight (165 tons).

T Roberts' design, *Rhadamanthus*, was, in 1833, the first British steamship, naval or mercantile, to cross the Atlantic, coaling in the Azores, but even so she relied on sail to a considerable extent. Though contemporary reports praise the *Rhadamanthus* she was converted to a troopship in 1841 while the others were on first-line duties much longer. The other two, *Salamander* (J Seaton) and *Phoenix* (Seppings) were described as poor sea boats. *Phoenix* was unique in that she was the only paddle warship to be converted to screw propulsion – at Woolwich in 1842-46. All four carried a heavy armament which was frequently altered but usually comprised two heavy pivots (10in or 8in) and two, later four, short 32-pounders.

Phoenix drew unfavourable comment in 1843. 'Very uneasy in a sea but from being strongly built does not steam much against a head sea and wind. Performance is very bad. Very slow and uneasy in bad weather but from great strength does not strain much. Would be a much handier vessel if rigged with three masts instead of two masts. Speed

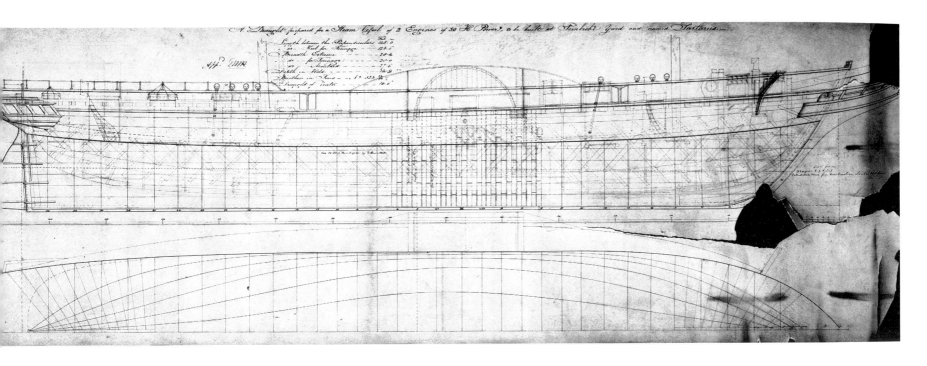

Rhadamanthus as built in 1833.

In 1833 she crossed the Atlantic, refuelling in the Azores. The lower paddle floats were removed for sailing, an unpleasant task for which stokers received an extra rum ration.

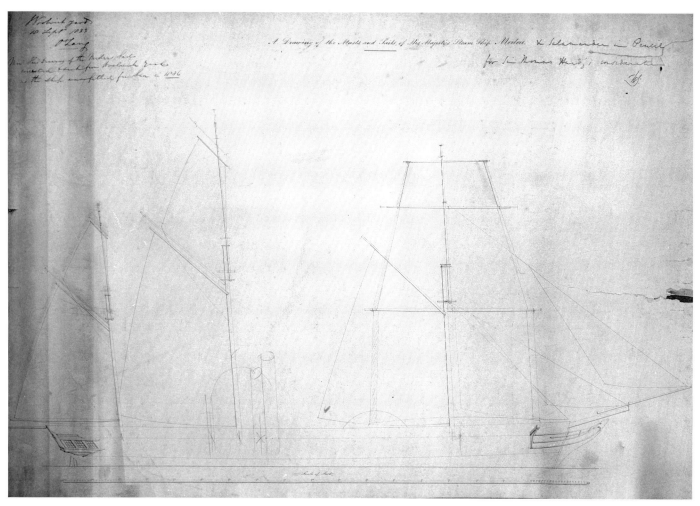

Medea, showing her modified rig with three masts in 1836.

The note in the top right says that the drawing, as modified in pencil, applies to *Salamander*. It appears that the funnel was moved slightly forward so that the main mast could be moved even further forward. The change from two to three masts is discussed in Chapter 11.

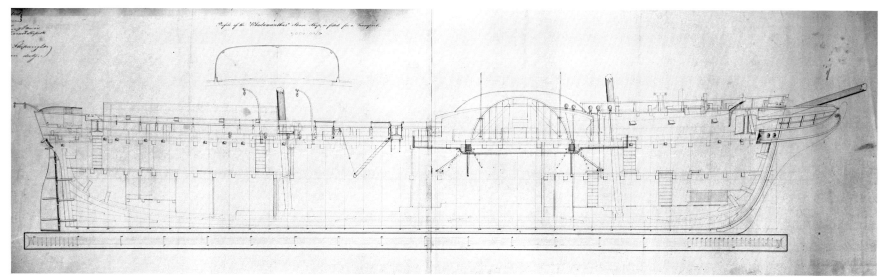

Rhadamanthus, the first British steamship to cross the Atlantic.

Her original configuration is shown in the photo on the previous page; this profile shows her after her modification in 1841. The crew lived in the high forecastle, leaving the lower deck to the troops. She has also been given paddle box boats.

under sail fair, under steam, very slow.'[11]

Far from being backward in the use of steamships as is often suggested, the Royal Navy led, both in numbers and in quality. A number of small countries, struggling for independence, had engaged the services of restless Royal Navy officers, several of whom ordered steam warships from British shipbuilders. Lord Cochrane ordered the *Rising Star* for Chile in 1821 though she arrived too late to fight, and Hastings had the famous *Karteria* and two smaller steamers built for Greece in 1826. The East India Company used the small *Diana* as an armed transport in the First Burmese War of 1824. She was probably the first steamship to go into action and she much impressed Royal Navy observers.

France had been interested in steamships since Henri-Joseph Paixhan's book of 1824 but did not complete a fighting ship, the *Sphinx*, until 1829. Even then, her engines had to be purchased from England, whilst Welsh coal was preferred. *Sphinx* took part in the French capture of Algiers in 1830. The US Navy used a hired vessel, *Seagull*, successfully against West Indian pirates in the 1820s.

Graham's Board was dedicated to cost cutting and this, together with the repercussions of the replacement of Seppings by William Symonds, no lover of steamships, led to a temporary slowing of building. One may note in passing the nine small ships acquired in the 1830s for service on the Great Lakes, to prevent gun running from the USA to Canada. Though not significant in the development of the steamship, they added to the experience of the Royal Navy in steam operation and further increased confidence in steamships. (Table 9)

Table 5: TRANSITIONAL SHIPS
Specification

Name	Tons bm	Disp't	Length ft in	Beam ft in	Depth ft in	Draught For'd ft in	Aft ft in
Medea	835	1142	179 4½	31 11	20 0	13 10	14 6
Rhadamanthus	813	1086	164 7	32 10	17 10	11 0	13 0
Salamander	818	1014	175 5	32 2	17 0	12 6	13 6
Phoenix	909	1024	174 7	31 10	16 9	12 0	12 6
Hermes	716	789	150 1½	32 9	18 2	–	–
Hermes (Lengthened)	830	–	170 0	32 9	18 2		
Volcano	720	1006	150 8 ⎫				
Acheron			150 0 ⎬ 32 9		18 0	11 6	12 0
Megæra			150 1 ⎭				
Hydra							
Hecate	817	1096	165 0	32 10	20 4	12 1	13 0
Hecla							
Alecto							
Ardent	800	878	164 0	32 8	18 7	11 3	11 8
Polyphemus							
Prometheus							

Notes:

Length is on the gun deck. Length overall is rarely quoted but, for example, *Medea* is 206ft oa.

Beam is that of the hull and is greater over paddle boxes, eg *Medea* 46ft oa.

Table 6: TRANSITIONAL SHIPS
Building Data

Name	Builder	Laid Down	Launched	Completed	Fate
Medea	Woolwich Dyd	Apr 1832	2 Sep 1833	1833	BU 1867
Rhadamanthus	Devonport Dyd	Sep 1831	16 Apr 1832	1832	BU 1864
Salamander	Sheerness Dyd	Apr 1831	16 May 1832	1832 at Woolwich	BU 1883
Phoenix	Chatham Dyd	May 1831	25 Sep 1832	1833	BU 1864
Hermes	Portsmouth Dyd	Apr 1834	26 Jun 1835	1836	BU 1864
Volcano	Portsmouth Dyd	Jul 1835	29 Jun 1836	1838	BU 1894
Acheron	Sheerness Dyd	1835	23 Aug 1835	1837	Sold 1855
Megæra	Sheerness Dyd	Aug 1836	17 Aug 1837	1837	4 Mar 1843, wrecked Bare Bush Key
Hydra	Chatham Dyd	Jan 1838	Jul 1838	1839	BU May 1870
Hecate	Chatham Dyd	Jun 1838	30 Mar 1839	1840	BU 1865
Hecla	Chatham Dyd	Jun 1838	14 Jan 1839	1839	BU Jun 1863
Alecto	Chatham Dyd	Jul 1839	7 Sep 1839	1839	BU 1865
Ardent	Chatham Dyd	Feb 1840	12 Feb 1841	1841	BU 1865
Polyphemus	Chatham Dyd	Feb 1840	28 Sep 1840	1841	29 Jan 1856, wrecked Jutland
Prometheus	Sheerness Dyd	Jul 1839	21 Sep 1839	1839	BU 1863

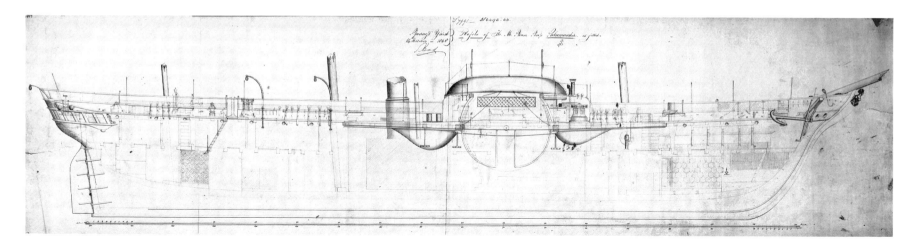

Salamander. This profile shows her in three masted style, ca 1840.

The paddle boxes have been extended to carry boats, providing useful space. Just visible ahead of the boat is what appears to be the galley chimney, keeping the smell and some of the heat out of the hull.

Table 7: GREAT LAKES PADDLE STEAMERS
Summary of Particulars

Name	Builder	Launch	Tons	Length ft in	Beam ft in	Depth ft in	NHP	Fate
Traveller	Niagara Dyd	1835	335	138 0	23 9	10 5	90	Sold 1844
Experiment	Niagara Dyd	1836	220	96 6	14 6	7 2½	25	Sold 1847
Montreal	Canada	1836	145	82 0	20 6	9 0	–	Deleted 1845
Toronto	USA	1834	342	147 0	29 4	7 10	–	Sold 1843
Canada	Niagara Dyd	Comp 1847	–	198 0	29 0	10 0	120	?
Minos	Chippawa	1840	406	152 8	24 2	13 2	90	Sold Mar 1852
Mohawk	Fairbairn	21 Feb 1843	174	99 1½	19 6	9 10	60	Sold Jun 1852
Sydenham	Montreal	1841	596	170 0	27 2	16 9	220	Sold Jul 1846
Cherokee	Canada	22 Sep 1842	751	170 0	30 10	16 3	200	Sold 1850

Notes

Traveller — Purchased 1835 at Niagara. 1x25pdr carronade. Crew 42.

Experiment — Built as sailing ship; purchased and converted 1838. 1x6pdr brass, 1x18pdr carronade. 10 men. Lake Huron 1839-46. Said to be an excellent sea boat (ADM 95-87).

Montreal — Acquired 1844. Schooner, 2 masts. (There is uncertainty as to whether she had engines) 1x18pdr. 35 men. Lake Erie.

Toronto — Built as *Adam*, bought 1838. 4x18pdr carronades.

Canada — Possibly stern wheeler. 2x68pdr.

Minos — Symonds. 1x12pdr. 29 men. Lake Erie.

Mohawk — Built in sections by Fairbairn, shipped to Canada and assembled at Kingston. Lengthened by 25ft in 1846. 1 gun. 20 men. Lake Ontario, later Huron.

Sydenham — Bought 27 Nov 1841 while under construction. Gun vessel, 1st class. Later packet in Mediterranean. 1x32pdr/26cwt, 2x32pdr/17cwt. (Possibly 2x68pdr later). 80 men.

Cherokee — Symonds, ex-brig sloop. 36 men.

3. Hermes and Derivatives

THE FIRST SHIP designed by Symonds and his Assistant Surveyor, John Edye, the *Hermes*, was approved in January 1834. As Surveyor, Symonds chose the form and approved the dimensions but Edye, a former Master Shipwright, carried out all the real design work. *Hermes* had Symonds' typical high rise of floor which made machinery installation more difficult and added to the draught.

She was said to be a 'good seaboat and behaves well under canvas but needs engines of additional power',12 but was otherwise not very satisfactory and in 1840 she was docked at Chatham to be lengthened by 20ft. Lang, who was in charge of the lengthening, said that it was 'to bring the bow up'. At the same time she was given a more powerful engine of Maudslay's Siamese design which increased her speed to 8½ knots.

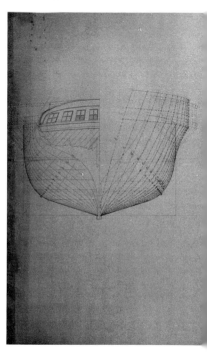

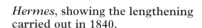
Hermes, showing the lengthening carried out in 1840.

Hermes was not liked in her original configuration; it is possible that she trimmed by the bow as Lang said that the lengthening was 'to bring the bow up'. Reports were still adverse but the lengthened form was the basis for a number of later ships. It may well have been a case of giving a dog a bad name, as almost identical ships received favourable comment.

Hermes, despite the lengthening, was still not very satisfactory as she steered badly. The contemporary explanation was that the thrust from the paddle wheels was now too far aft, a viewpoint which led to opposition to the screw propeller, thrusting right aft. With hindsight, one cannot accept this explanation for *Hermes*' defect but there is insufficient evidence to hazard a guess as to the real cause. *Hermes* was built as a brig but, as with most steam vessels, she was later given a mizzen.

Hermes was used as the basis for several other classes even though she herself was generally regarded as unsatisfactory. The conflict between criticism of *Hermes* and praise for almost identical ships shows the unreliability of subjective reporting. However, criticism of Symonds' designs must be treated with care as his opponents were almost as dogmatic as he was. Three almost identical ships of the *Volcano* class were also ordered in 1834. *Megæra* was wrecked in 1843 and the other two were reclassified as sloops, third class, in 1845. *Volcano* was converted to a floating 'factory' – fleet repair ship – in 1854. Her captain wrote around 1840 that she was 'very buoyant and easy, rolls deep and pitches easily in a head sea' but that she needed more power. [13]

These vessels were followed by the four very similar ships of the *Alecto* class, ordered in 1839. A fifth ship, *Rattler*, was re-ordered and modified as the prototype screw warship while two of her half sisters, *Polyphemus* and *Alecto*, were used in comparative trials with the screw ship. All these *Hermes* derivatives carried one or two pivot guns (68-pounders, 8in or 10in) and two, later increased to four, short 32-pounders though the outfit was altered frequently. The *Alecto*s were reclassed as sloops, third class. During the Parana river campaign of 1843, *Alecto* was given extra ports so that both her 32-pounders could fire on either side. She also mounted two stands of sea service rockets.

She was described as 'very easy and buoyant both steaming and under canvas and steers remarkably well'. [14]

The three *Hydra*s are usually described as based on the lengthened

Table 8: TRANSITIONAL SHIPS
Machinery Particulars

Ship	Engines by	NHP	Type	No Cyl	iph=	speed (trials)
Medea	Maudslay	220	Side lever	2		
1846	Maudslay	220	Siamese	4	900	10.6
Rhadamanthus	Maudslay	220	Side lever	2	385	10.0
Salamander	Maudslay	220	Side lever	2	506	7.2
Phoenix	Maudslay	220	Side lever	2		
Hermes	Maudslay	220	Side lever	2		
1842	Maudslay	220	Siamese	4	–	8.5
Volcano	Seaward	140	Side lever	2		
Acheron	Seaward	160	Side lever	2		
Megæra	Seaward	140	Side lever	2		
Hydra	Watt	220	Side lever	2		c9
Hecate	Scott	240	Side lever	2		c9
Hecla	Scott	240	Side lever	2		c9
Alecto, Ardent, Polyphemus and *Prometheus*	Seaward	200	Direct	2	370	8.5/9.5kts

Notes

The variation in speed between ships of the same class illustrates the inaccuracy of speed measurement in these early days, the meaninglessness of NHP and, possibly, some difficulty in measuring ihp.

Builders' names have been abbreviated in this and later tables from Seaward & Capel to Seaward; Boulton & Watt to Watt; Scott & Sinclair to Scott.

Most engines were installed at Woolwich Dockyard (*Alecto* at Limehouse).

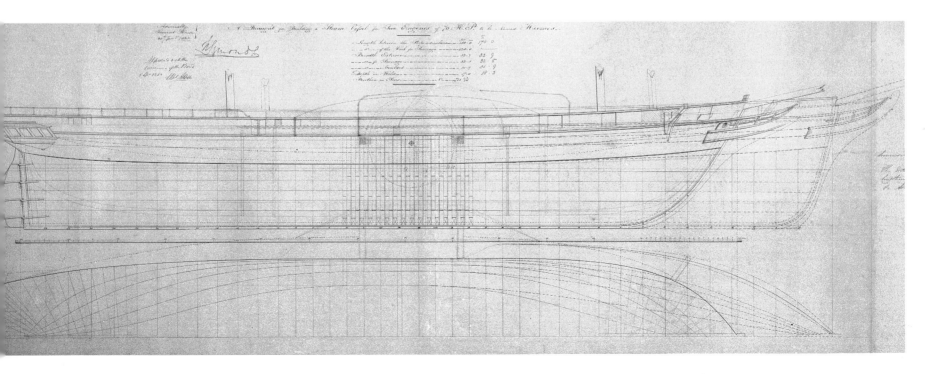

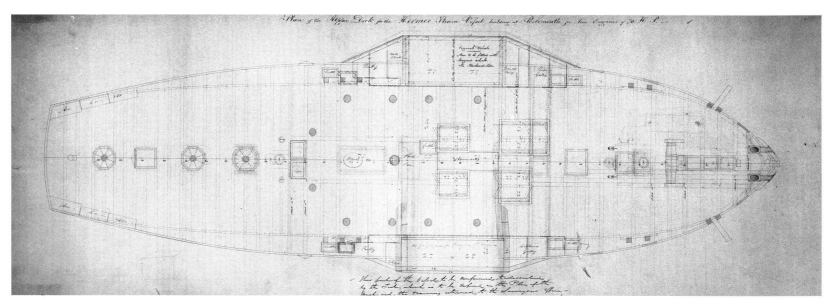

Hermes.

This upper deck plan *(above)* shows how much space was available, facilitating frequent changes of armament. The numerous small circles amidships are coaling hatches.

Volcano as a 'Floating Factory' (repair ship) in 1854.

An auxiliary engine aft works a line of shafting extending under the upper deck for most of the length, from which belts drive lathes, etc. There is a steam hammer and what is probably a small blast furnace. The officers have moved up to a small deck house aft but there can have been little room for the crew.

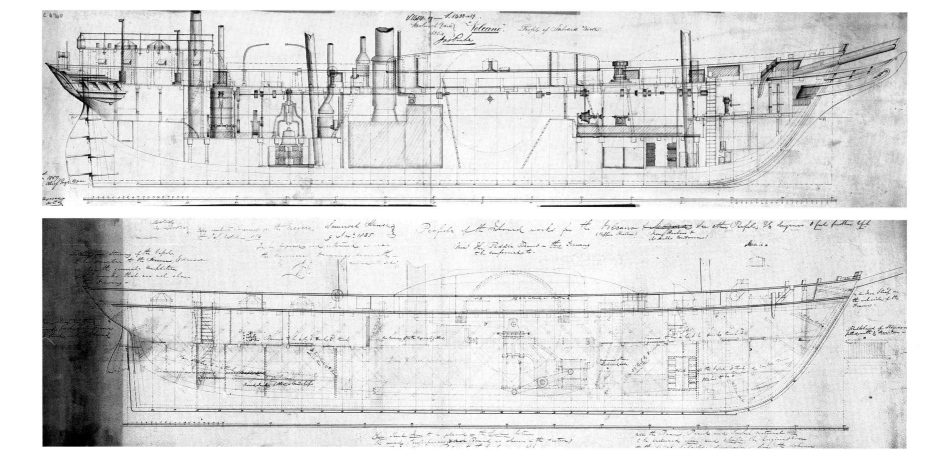

Hecate aground in 1861.

In this drawing, the artist gives an unusual deck view of the ship.

Table 9: TRANSITIONAL WARSHIPS
Armament

Where the date of a change in armament is not known, the outfits will be numbered 1, 2 etc. In these early ships, all guns were carried on the upper deck.

Ship	Date	Armament
Medea	1842	2x10in/84cwt pivot. 2x32pdr/24cwt
	1856	2x10in/84cwt pivot. 4x32pdr/25cwt
	1862	1x110pdr/82 RBL pivot, 1x10in/84cwt pivot, 4x32pdr/25cwt
Rhadamanthus	1	1x10in/84cwt pivot, 2x32pdr/25cwt, 2x6pdr/6cwt brass
	2	1x10in/84cwt pivot, 1x32pdr/42cwt pivot, 2x32pdr/25cwt
	As transport	2x18pdr/22cwt
Salamander	1842	2x10in/84cwt pivot, 2x32pdr/25cwt
	1856	2x10in/84cwt pivot, 4x32pdr/25cwt
	1862	1x110pdr/82 RBL pivot, 1x10in/84cwt pivot, 4x32pdr/25cwt
Phoenix		1x10in/84cwt pivot, 1x8in/52cwt pivot, 4x32pdr/17cwt
Hermes	As completed	2x9pdr/13.5cwt brass
	1843	1x8in/52cwt pivot, 2x32pdr/17cwt
	3	2x8in/52cwt pivot, 2x32pdr/25cwt,
	1856	2x8in/52cwt pivot, 4x32pdr/25cwt,

Ship	Date	Armament
	1862	2x40pdr/35cwt RBL pivot, 4x32pdr/25cwt
Volcano	All as comp	2x9pdr/13.5cwt brass
	1842	1x68pdr/95cwt pivot, 4x32pdr/17cwt
	1856	1x32pdr/42cwt pivot, 2x32pdr/25cwt
Acheron		
Megæra		
Hydra OR	As completed (accounts differ)	2x68pdr/95cwt pivot, 2x32pdr/25cwt
		2x8in/65cwt pivot, 2x32pdr/50cwt,
	1842-51	2x10in/84cwt pivot, 2x32pdr/34cwt,
	1856	2x10in/84cwt pivot, 4x32pdr/25cwt
	1862	1x110pdr/82 RBL pivot, 1x10in/84cwt pivot, 4x32pdr/17cwt
Hecate		
Hecla		
Alecto	As completed	2x32pdr/45cwt pivot, 2x32pdr/25cwt
	1841-48	1x68pdr/95cwt pivot, 2x32pdr/17cwt
	1850	1x68pdr/95cwt pivot, 4x32pdr/17cwt
	1856	1x32pdr/45cwt pivot, 4x32pdr/25cwt
	1862	1x40pdr/35cwt RBL pivot, 2x32pdr/25cwt
Ardent/Polyphemus		
Prometheus		

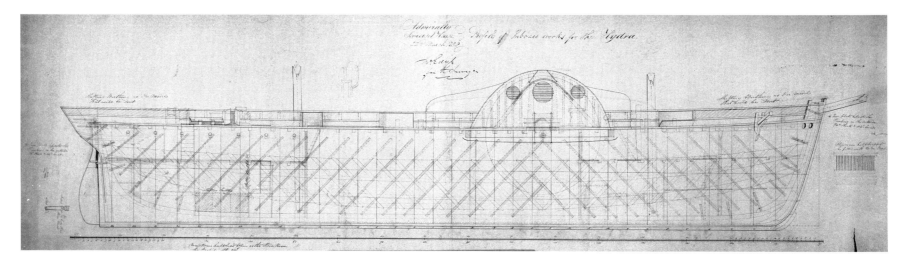

Hydra as designed in 1837.

She was typical of the smaller sloops of the period.

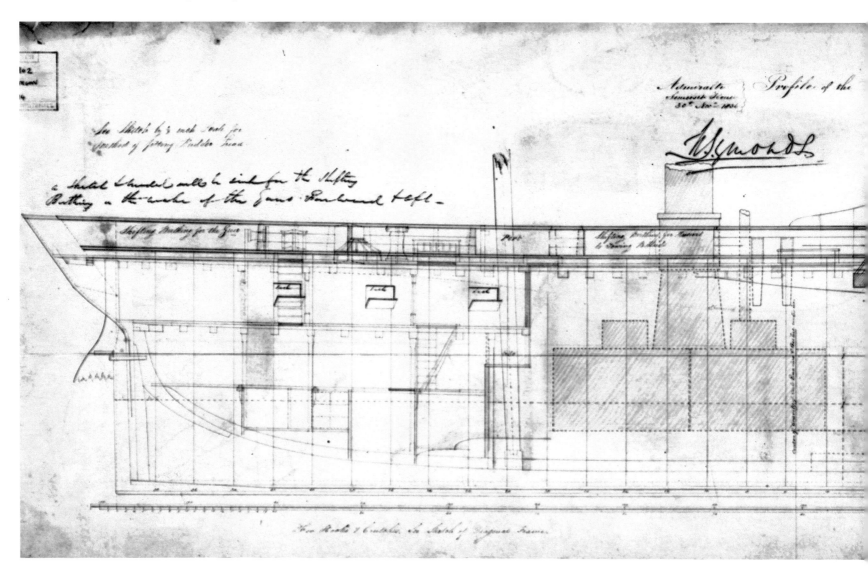

Hermes, which cannot be literally true since they were ordered in 1837-38. It may be that it was already intended to lengthen *Hermes* by then. Their armament was much the same as the *Alecto*s though they usually mounted two pivot guns. Since the *Hydra*s were much overweight as completed the guns were usually landed for a long voyage. They had coal for 2,400 miles at full power (ten days) and were said to be good sea boats.

At about £25,000 these ships were not expensive though, as Dockyard-built ships, overheads were not treated the same way as in commercial yards.[15] It should not be forgotten that they cost twice as much as a sailing ship of the same size.

Gorgon (1835) was bigger, faster and more heavily armed than earlier ships.

John Edye's design was seen as a great success and 28 more generally similar ships were built, either as sloops or frigates. The drawing shows the main deck ports in which guns were never mounted. Note also the simpler engine framing compared with that of the earlier, side lever ships.

4. Gorgon and Cyclops and their Derivatives

ALMOST BY CHANCE, the *Gorgon* marked a major advance in the design of paddle warships and was the prototype for a further twenty-eight ships. At the end of 1833, Symonds and Edye began the design of what was at first described as a slightly enlarged *Medea* with 220NHP machinery. For *Gorgon*, Seaward & Capel offered engines of 350NHP and an increased coal stowage of 380 tons within the limits set. This tender was accepted, thereby making *Gorgon* a much more powerful and faster ship than previous steam vessels.

The engine design, which became known as the *Gorgon* engine and was used in several later ships, used a short connecting rod from the piston to the crankshaft which, working at awkward angles, led to heavy

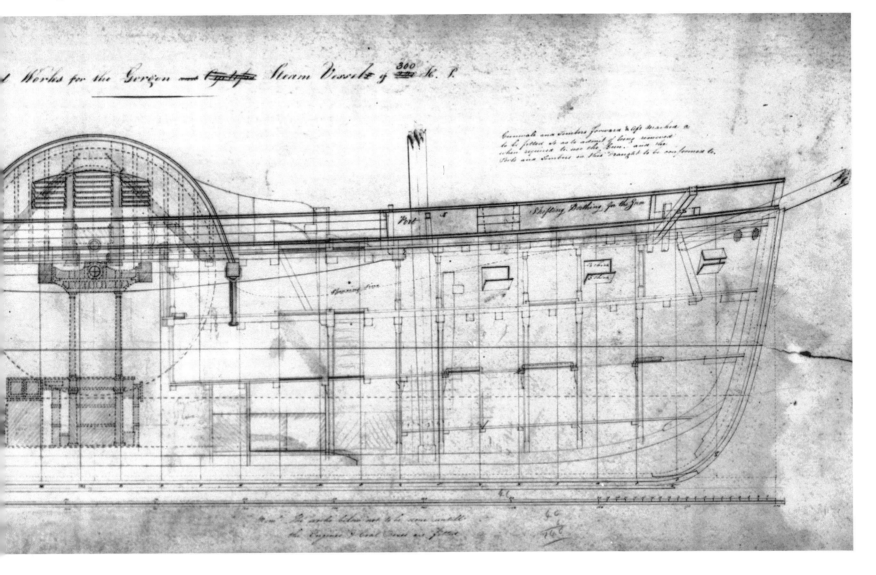

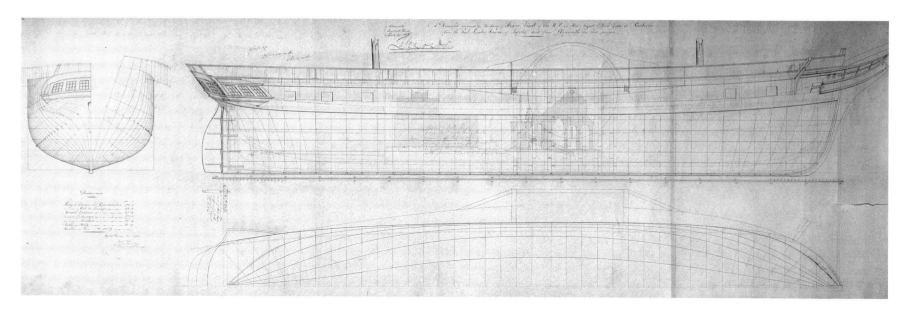

Lines plan of *Gorgon*.

Symond's influence is apparent in the high rise of floor shown in the sections.

Table 10: GORGON AND CYCLOPS
Specification

Name	Tons bm	Disp't	Length ft in	Beam ft in	Depth ft in	Draught For'd ft in	Aft ft in
Gorgon	1111	1610	178 0	37 6½	23 0	16 2	17 3
Cyclops	1195	1960	190 3	37 6	23 0	–	17 10

Table 11: GORGON AND CYCLOPS
Building Data

Name	Builder	Laid Down	Launched	Completed	Fate
Gorgon	Pembroke Dyd	July 1836	31 Aug 1837	1838 at Deptford	BU 1864
Cyclops	Pembroke Dyd	Aug 1838	10 July 1839	1839	BU Jan 1864

Table 12: GORGON AND CYCLOPS
Machinery Particulars

Ship	Engines by	NHP	Type	No Cyl	Iph = speed (trials)
Gorgon	Seaward	320	'Gorgon'	2	800 9.5
Cyclops	Seaward	320	'Gorgon'	2	1100 9.5

Table 13: GORGON AND CYCLOPS
Armament

Ship	Date	Armament
Gorgon	1838	2x42pdr/84cwt pivot, 2x68pdr/64cwt, 2x42pdr/22cwt
	1856	1x68pdr/95cwt pivot, 1x10in/84cwt pivot, 4x32pdr/42cwt
Cyclops	as comp.	2x98pdr/?? cwt, 4x68pdr/95cwt
	1845	2x10in/84cwt pivot, 4x8in/65cwt
	1856	2x68pdr/95cwt pivot, 4x10in/84cwt

vibration and severe wear on the bearings. Steam was supplied at 5lb/sq in from four tubular boilers. The coal bunkers were 8ft wide, outboard of the machinery, providing considerable protection when full. The machinery cost £21,073 and drove common wheels 27ft in diameter with forty-eight teak floats, 7ft wide, at 18rpm, giving 9.8 knots on trials. These were later replaced by cycloidal wheels.

Gorgon had teak frames, cut in Bombay many years earlier for the frigate *Tigris*. She was a big ship for the day with a displacement of 1610 tons, only a little smaller than Brunel's *Great Western* of 2372 tons. The structure of both these ships used Edye's modification of Sepping's diagonal system.[16] *Gorgon* cost £54,301. She was a handsome, brig-rigged vessel with a single funnel abaft the paddle boxes and was favourably reported on as a good sea boat. Her first captain, W H Henderson, said 'I consider the vessel powerful, easy and a first rate sea boat and one, when not too deep, of more than average speed under steam'.[17] She was also good under sail, at least in calm water. Henderson also said that when steaming into a head wind (he seems to have meant about 20 knots wind speed) she was 2-3 knots slower under full rig than with the topmasts housed, and that typical fouling of the copper-sheathed bottom would reduce her speed by 1½ knots.

The hull was spacious and she could carry 1600 troops together with six field guns and their limbers. Her design armament is discussed with that of *Cyclops*.

It was decided in 1836, before *Gorgon* was launched, to order a sister ship to be called *Cyclops*. She was not laid down until 1838, by which time it had been decided to lengthen her by 12ft 3in. On completion, she was rated as a frigate which is surprising since all her armament was on the upper deck. Frigates were then defined as ship-rigged vessels, and *Cyclops* was a brig, with the principal battery on the main deck

Gorgon, **shown lying in the Hamoaze, Plymouth.**

This fine lithograph shows the main deck gunports which were never used, and shows that they were quite high out of the water. The immersion of the paddle wheels seems correct for normal draught.
(The Science Museum, London)

covered by a complete upper deck. She was of familiar style, with circular paddle boxes, a knee bow and a square stern.

Symonds' biographer, Sharpe[18], gives the following design armament for the two ships.

Table 14: Design and Fitted Armament

Ship	Upper deck	Main deck	As completed. Upper deck only
Gorgon	6x32pdr	10x32pdr	2x42 (pivot), 2x68, 2x42pdr
Cyclops	2x98, 4x48pdr	16x32pdr	2x98 (pivot), 4x68pdr

The builder's model of *Gorgon*, held in the Science Museum, London, shows an arrangement of gunports consistent with Sharpe's figures[19]. Edye, in evidence to the 1847 Select Committee, says that the two ships were not designed to carry guns on the main deck but that the upper deck guns could be lowered to that deck for use in riverine fighting and hence ports were provided, lightly caulked, and ring bolts fitted. This story is not consistent with the number of main deck ports in either ship and it is interesting to note that when *Gorgon* was fighting in the Parana river campaign the guns were not lowered.

It was said at the time that the ships, particularly with their machinery, were overweight and that the loss of freeboard made it impossible to carry main deck guns. *Gorgon*'s machinery was 32 tons overweight, reducing freeboard by 3in, and a further 18in would have been lost if the extra weight of coal was not allowed for in the design. The draughts given in Table 10 are much greater than those of later ships with similar

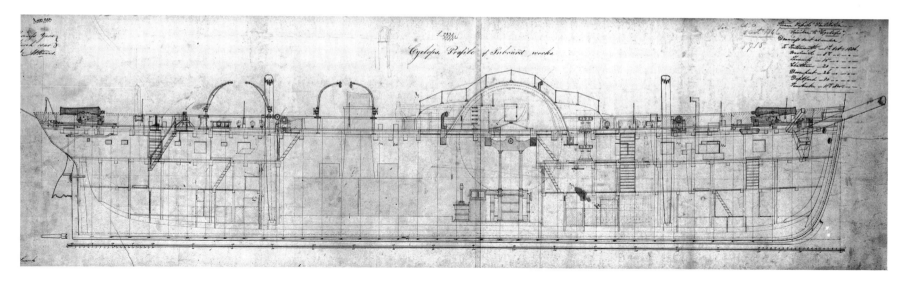

Cyclops, a second class frigate.

She was only slightly longer than *Gorgon* and carried a fairly similar armament, so the reason for the difference in classification is not clear. The guns seen at bow and stern were 98 pounders as completed.

depth but illustrations suggest that both ships had generous freeboard and this is quoted as 6ft 4in to the main deck, though it is not clear if this is a design figure or as completed. It is pertinent to note that the only *Gorgon* derivatives to carry a main deck armament were 18in deeper in the hull. However, *Gorgon*'s freeboard seems adequate as completed and there was plenty of room for the larger number of gun crews required. As with all ships of the day, the armament was changed frequently but always comprised two heavy pivots, fore and aft, and two more heavy guns on each broadside. (See Table 13)

Despite this apparent problem with the armament, both ships were successful in service and many derivatives were built, as sloops, first class, based on *Gorgon*, and as frigates, second class, based upon *Cyclops*.

Driver, a first class sloop derived from Gorgon.

Though Symonds was still Surveyor when *Driver* was ordered in 1841, his authority was much reduced and John Edye was able to give her a more conventional form with less rise of floor.

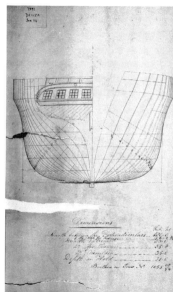

Driver.

This profile shows that the lower deck ports have been omitted. Note also that, though the paddle sponsons are growing in size, they do not carry boats.

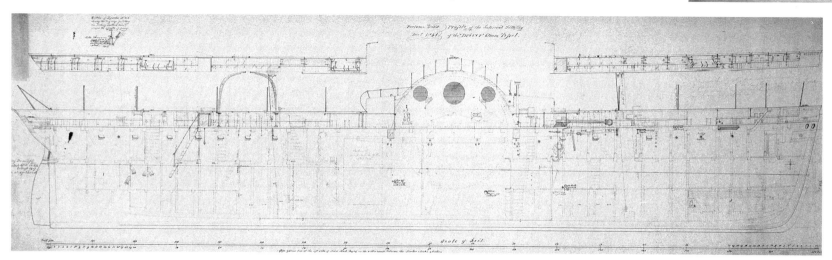

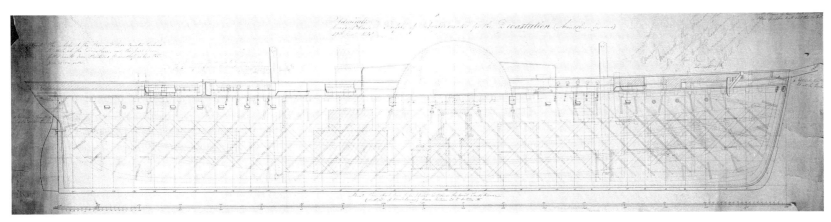

Devastation, a variant of the *Driver* class.

She had a more powerful Maudslay engine which gave her a trial speed of about 10 knots. The cylinders had a stroke of 72 inches and a diameter of 54 inches and used steam at 5½ lb/in².

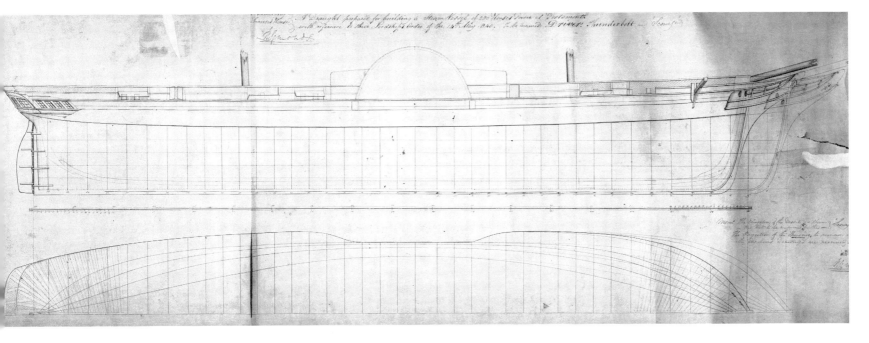

Gorgon derivatives, sloops, first class

Gorgon had the high rise of floor typical of Symonds; a feature much reduced in later ships, partly due to Symonds' diminishing authority. The first two, *Stromboli* and *Versuvius*, were ordered in 1838, still as steam vessels of the *Medea* class. All steam vessels were reclassified in 1844 when these ships became sloops, first class, with a complement of a commander, three lieutenants and 145 men. They had much reduced rise of floor and were slightly smaller than *Gorgon* with less beam and draught. Their armament was similar to that of *Gorgon* and there were no main deck ports.

The eight ships of the *Driver* class were approved in February 1840 'as *Stromboli*' but were re-ordered in March 1841 to a modified design. Though there are references in Admiralty files to order dates, these seem to have been an almost irrelevant formality since, in several cases,

building was already well advanced. The beam was increased, though not back to *Gorgon*'s figure, to permit bigger bunkers. They cost between £36,000 and £50,000 and, like most early steam vessels, were completed as brigs or brigantines but soon given a mizzen, so becoming barques.

Devastation was very similar but with more powerful Maudslay Siamese engines of 400NHP which on trial developed 785ihp for a speed of 10 knots. She was laid down in 1840, as were the earlier *Driver*s, and is best regarded as an experimental variant of that class.

Two more, *Thunderbolt* and *Virago*, were laid down in 1841 with 300NHP engines. Two further ships with more design changes were approved in March 1841 but, due to a shortage of building slips, they were not laid down until 1844 which allowed time for more changes. The first was *Sphynx* (later renamed *Sphinx*) in which the engine room

Table 15: *GORGON* DERIVATIVES – FIRST CLASS SLOOPS
Building Data

Name	Builder	Laid Down	Launched	Completed	Fate
Stromboli	Portsmouth Dyd	Sep 1838	27 Aug 1839	1840 Portsmouth	BU 1867
Vesuvius	Sheerness Dyd	Sep 1838	11 Jul 1839	1840 Chatham	BU 1866
Cormorant	Sheerness Dyd	7 May 1841	29 Mar 1842	1843	BU Aug 1853
Driver	Portsmouth Dyd	June 1840	24 Dec 1840	1841	Wrecked Aug 1861
Geyser	Pembroke Dyd	Aug 1840	6 Apr 1841	1842	BU Jul 1876
Growler	Chatham Dyd	Jan 1841	20 Jul 1841	1841	BU Jan 1854
Eclair, ex-*Infernal*	Woolwich Dyd	Aug 1841	31 May 1843	1844	BU 1864
Spiteful	Pembroke Dyd	Aug 1841	24 Mar 1842	1842	BU 1883
Styx	Sheerness Dyd	22 Jun 1840	26 Jan 1841	1841	BU Apr 1866
Vixen	Pembroke Dyd	June 1840	4 Feb 1841	1842	BU 1862
Devastation	Woolwich Dyd	27 Jul 1840	3 Jul 1841	1841	BU 1867
Thunderbolt	Portsmouth Dyd	Apr 1841	13 Jan 1842	1842 Pembroke	Wrecked 3 Feb 1847
Virago	Chatham Dyd	15 Nov 1841	25 Jul 1842	1843 Woolwich	BU Nov 1876
Sphynx	Woolwich Dyd	May 1844	17 Feb 1846	1846 Woolwich	BU 1881
Bulldog	Chatham Dyd	7 Jul 1844	2 Oct 1845	1846 at Devonport	23 Oct 1865, burnt off Haiti
Fury	Sheerness Dyd	Jun 1845	31 Dec 1845	1847 Sheerness	BU Jul 1864
Inflexible	Pembroke Dyd	Jan 1844	24 May 1845	1846	BU Jul 1864
Scourge	Portsmouth Dyd	Feb 1844	9 Nov 1844	1845 East India Dock	BU 1865
Basilisk	Woolwich Dyd	Nov 1846	22 Aug 1848		BU 1882

Table 16: *GORGON* DERIVATIVES – FIRST CLASS SLOOPS
Specification

Name	Tons bm	Disp't	Length ft in	Beam ft in	Depth ft in	Draught For'd ft in	Aft ft in
Stromboli	970	1283	180 0	34 5	21 0	13 0	13 6
Vesuvius							
Cormorant	1054-1059	1379	180 0	36 0	21 0	13 2	14 7
Driver		(light)	203 9oa	45 0oa			
Geyser							
Growler							
Eclair							
Spiteful							
Styx							
Vixen							
Devastation	1058	1380 (light)	180 0	36 0	21 0	14 3	14 9
Thunderbolt	1059	1669	180 0	36 0	21 –	14 3	14 9
Virago							
Sphynx	1056	1611	180 0	36 0	20 11	–	–
Bulldog	1124	–	190 0	36 0	21 0	–	–
Fury				57 4oa			
Inflexible							
Scourge							
Basilisk	1031	1710	190 0	34 5	21 5	15 4	16 6

Table 17: *GORGON* DERIVATIVES – FIRST CLASS SLOOPS
Machinery particulars

Ship	Engines by	NHP	Type	No Cyl	ihp =	speed (trials)
Stromboli and *Vesuvius*	Napier	200	Side lever	2		c9
Cormorant	Fairbairn	280	Direct	2	–	–
Driver	Scott Sinclair	280	Direct	2		
Geyser	Scott Sinclair	280	Direct	2		
Growler	Scott Sinclair	280	Direct	2		
Eclair	Miller	280	Direct	2		
Spiteful	Scott Sinclair	280	Side lever	2		
Styx	Scott Sinclair	280	Direct	2		
Vixen	Scott Sinclair	280	Direct	2		
Devastation	Maudslay	400	Siamese	4	785	10
Thunderbolt	Napier	300	Direct	2	–	10
Virago	Watt	300	Direct	2	–	10
Sphynx	Penn	500	Oscillating	2	–	12
Bulldog	Rennie	500	Direct	2	–	10.2
Fury	Rigby	515	Direct	2	–	10.5
Inflexible	Fawcett	378	Direct	2	680	9.5
Scourge	Maudslay	420	Direct	2	–	11
Basilisk	Miller	400	Oscillating	2	1033	–

was lengthened 6ft without changing the hull. Tenders were invited for the most powerful machinery, a competition won by Penn with oscillating engines of 500NHP at a cost of £25,000. The hull also cost £25,000. It is claimed that she reached 12 knots on trial, though this may be doubted since particulars of the trial are not given and none of the next group with engines of similar size exceeded 11 knots and most were slower.

The four ships of the *Bulldog* class had the lengthened engine room of *Sphynx* but in a hull 10ft longer. The engines were by four different builders and NHP ranged from 378 to 515, the Fawcett engines of 378NHP delivering 680ihp on trial for a speed of 9.5 knots, the slowest of the class. *Scourge* had two funnels close together behind the paddle boxes, the others had a single funnel. Armament was as *Gorgon*.

Though not a *Gorgon* derivative, one may conclude this section with mention of *Basilisk*, which was built to the lines of Lang's screw sloop *Niger* for comparative trials, discussed later. Her boilers were arranged fore and aft of a fairly short engine room, permitting the main mast to be stepped ahead of the second funnel, and the mizzen was further forward than in other paddle sloops. This balanced rig made her the best sailer of the category.

Cyclops derivatives, frigates, second class

The distinction between a first class sloop and a second class frigate was always unclear, the armament being quite similar, and some frigates were reclassified as sloops late in life. Including *Cyclops*, there were seven very similar frigates and three rather more distant derivatives. The first seven were built as brigs but soon became barques. *Cyclops* had circular paddle boxes but the others had boxes adapted to carry large boats.

The first six were all approved in March 1841 and tenders for their engines were invited in November. The last three were delayed, possibly due to a shortage of slips, and a number of changes were made to the design. The three ships of the *Firebrand* class had bigger engines than *Cyclops* (400-470NHP) but their recorded speeds of 9.5 to 10 knots

Table 18: *GORGON* DERIVATIVES – FIRST CLASS SLOOPS
Armament

Ship	Date	Armament
Stromboli	As comp	2x42pdr/84cwt pivot, 2x68pdr/64cwt, 2x42pdr/22cwt
Vesuvius	1856	1x68pdr/95cwt pivot, 1x10in/84cwt, 4x32pdr/42cwt
	1856 (2)	1x10in/84cwt pivot, 1x110pdr/82cwt RBL, 4x32pdr/42cwt
Cormorant	As comp	2x42pdr/84cwt pivot, 2x68pdr/64cwt, 2x42pdr/22cwt
Driver	1856	1x68pdr/95cwt pivot, 1x10in/84cwt, 4x32pdr/42cwt
Geyser		(*Geyser* only main deck 1x42pdr/75cwt)
Growler	1862	1x10in/84cwt pivot, 1x110pdr/82cwt RBL, 4x32pdr/42cwt
Eclair		(*Geyser* (1861), *Spiteful* (1862) 4x8in/52cwt in place of 32pdr)
Spiteful		(*Driver* 1x110pdr in place of 68pdr)
Styx		
Vixen		
Devastation	1842	2x42pdr/84cwt pivot, 2x68pdr/64cwt, 2x42pdr/22cwt
	1856	1x68pdr/95cwt pivot, 1x10in/84cwt, 4x32pdr/42cwt
	1862	1x10in/84cwt pivot, 1x110pdr/82cwt RBL, 4x32pdr/42cwt
Thunderbolt	As comp	2x42pdr/84cwt pivot, 2x68pdr/64cwt, 2x42pdr/22cwt
Virago	1856	1x68pdr/95cwt pivot, 1x10in/84cwt, 4x32pdr/42cwt
	1861	1x10in/84cwt pivot, 1x110pdr/82cwt RBL, 4x32pdr/42cwt
	1865	5x64pdr/58cwt MLR, possibly 1x100pdr/9in/6¼ton SB
Sphynx	As *Virago*	
Bulldog	As comp	2x42pdr/84cwt pivot, 2x68pdr/64cwt, 2x42pdr/22cwt
Fury	1856	1x68pdr/95cwt pivot, 1x10in/84cwt, 4x32pdr/42cwt
Inflexible	1862	1x10in/84cwt pivot, 1x110pdr/82cwt RBL, 4x32pdr/42cwt
Scourge		(*Inflexible* 4x8in/52)
Basilisk	1842	2x42pdr/84cwt pivot, 2x68pdr/64cwt, 2x42pdr/22cwt
	1856	1x68pdr/95cwt pivot, 1x10in/84cwt, 4x32pdr/42cwt
	1862	1x10in/84cwt pivot, 1x110pdr/82cwt RBL, 4x32pdr/42cwt
	Later	5x64pdr/71cwt MLR

Stromboli (1838), a first class sloop based on *Gorgon*.

She had a little less power and less beam than *Gorgon* and had much reduced rise of floor. Such ships were commanded by a commander with three lieutenants and 145 men.

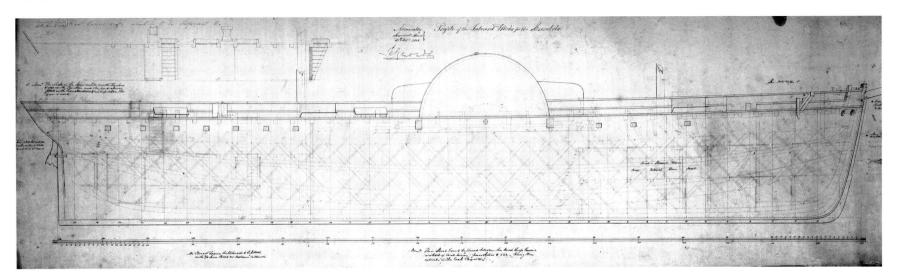

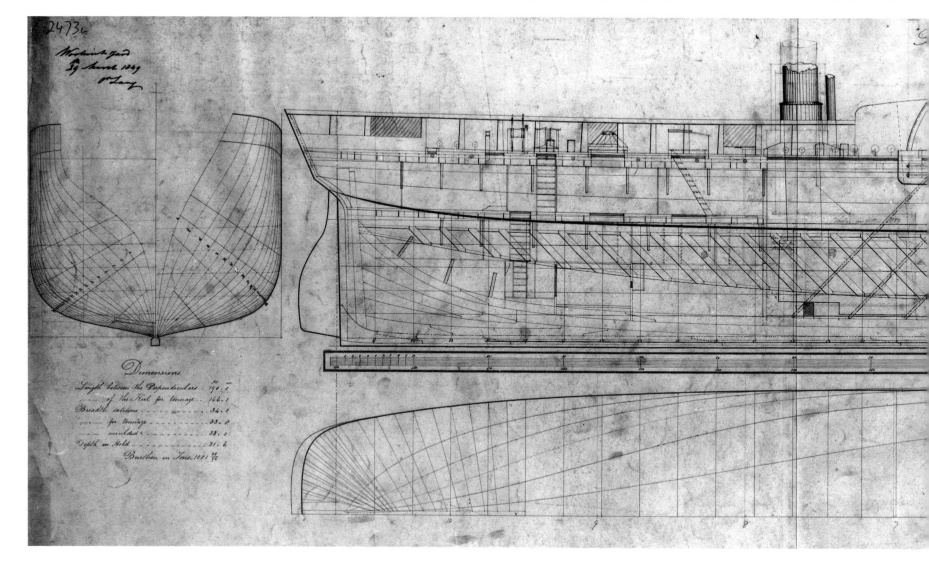

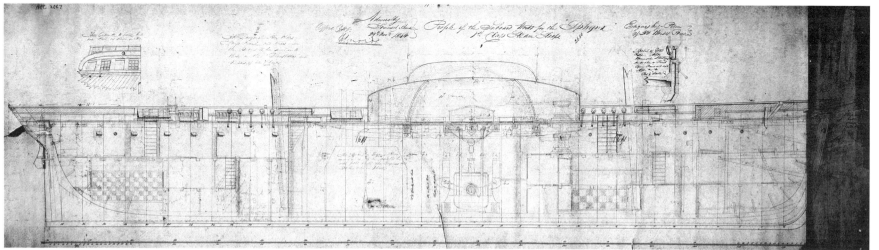

Sphynx, another variant of the first class sloop.

Her engine room was lengthened by 6ft to permit the fitting of more powerful engines. It is claimed that she reached 12 knots on trial though this seems unlikely, except with the tide.

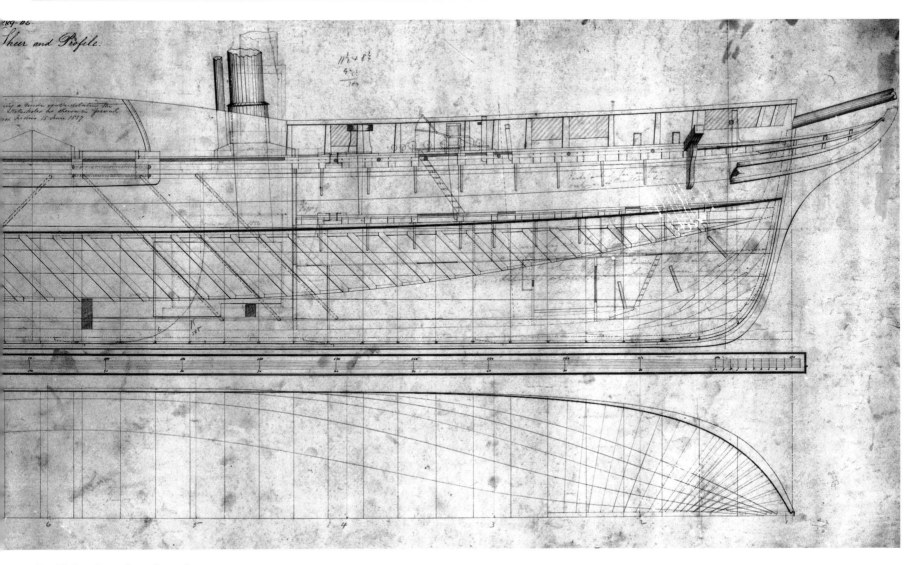

Sheer and Profile.

Basilisk, a later first class sloop.

She was built to be as similar as possible to Lang's screw sloop *Niger* for comparative trials. The lighter hull and machinery of the screw ship made the answer clear.

were little different. *Firebrand*, only, had boilers working at 8lb/sq in instead of the 5 in earlier ships. *Vulture* had twin funnels, side by side, whilst the others had a single funnel. They all had paddle box boats.

Sampson seems to have been the next design even though she was laid down after the following two. She had much less rise of floor and was 13ft 6in longer. Her 467NHP engines drove wheels of 27ft 6in diameter giving her a speed of 9.5 knots. The remaining two ships had the same hull as *Sampson* but with higher-pressure boilers (8-10lb/sq in) and engines of 540NHP (*Centaur*) giving 9.5 knots and 560NPH (*Dragon*) giving 11.5 knots.

Tiger was designed by Edye and approved in February 1847 as a sloop of the *Sphynx* class. In June that year her depth of hold was increased by 18in and it is said that it was the extra tonnage which led to her reclassification as a frigate, but her extra depth also made it possible to install guns on the main deck, another 'frigate' characteristic.

She carried the usual armament on the upper deck but, at last, was given a main deck battery, initially four, later ten, 32-pounders, and this would certainly have justified frigate classification.

She was heavily built with two layers of planks, each 4-6in thick and had a very flat floor. *Tiger* was the biggest ship lost by the Royal Navy during the Crimean War when she ran aground near Odessa in fog, and was set on fire by shore batteries.

Magicienne and *Valorous* were also approved in February 1847 as *Sphynx* class sloops but in August of that year it was decided to build them to a design of Edye's similar to *Tiger* but 5ft longer. Their twin funnels were side by side. *Valorous* was the last major paddle fighting ship completed for the Royal Navy, excluding minesweepers of World War I, and was the last to remain in active service, only being scrapped in 1891.

Furious was originally approved as a *Sphynx* class sloop but the requirements were greatly altered and new plans by Fincham were approved in February 1847. She carried two pivot guns and four 32-pounders on the upper deck and eight to ten 32-pounders on the main deck.

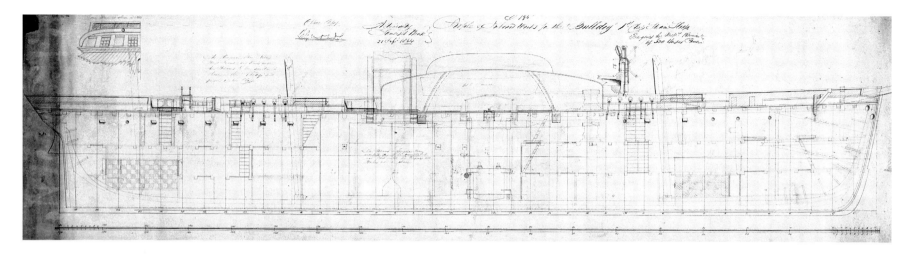

Table 19: *CYCLOPS* DERIVATIVES – SECOND CLASS FRIGATES
Specification

Name	Tons bm	Disp't	Length ft in		Beam ft in		Depth ft in		Draught For'd ft in	Aft ft in
Firebrand, Gladiator and *Vulture*	1190	1960	190	0½	37	6	23	0	–	–
Sampson	1299	c2100	203	6	37	6	23	0	–	–
Centaur Dragon	1270	c2100	200	0	37	6	23	0	–	–
Avenger	1444	–	210	0	39 65	0 6oa	25	8	–	–
Other Second Class Frigates										
Birkenhead	1405	1918	210	0	37 60	8 6oa	22	11	15 9 (mean)	
Tiger	1221	–	205	0	36	0	24	6	17 0 (mean)	
Magicienne Valorous	1255	2300	210	0	36	0	24	6	–	–
Furious	1287	–	206	0	36	6	23	3	–	–

Bulldog was the last of the *Gorgon* derivatives.
She was 10 feet longer than *Sphynx*.

Firebrand, a second class frigate derived from *Cyclops*.

There was little difference between the second class frigate and the first class sloop. Though this drawing shows the lower deck gun ports, she never carried guns in them.

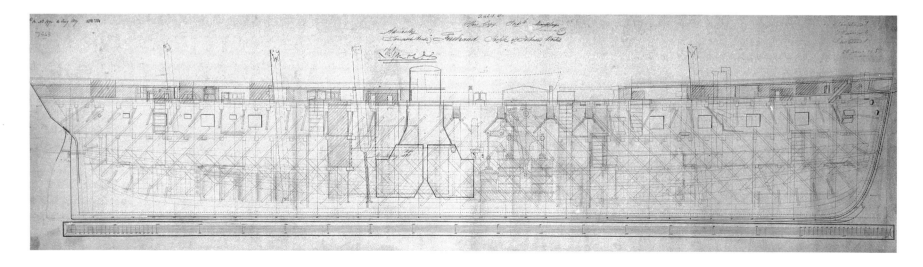

Table 20: *CYCLOPS* DERIVATIVES – SECOND CLASS FRIGATES
Building Data

Name	Builder	Laid Down	Launched	Completed	Fate
Firebrand	Portsmouth Dyd	Dec 1841	5 Sep 1842	June 1844	BU Oct 1864
Gladiator	Woolwich Dyd	Feb 1842	15 Oct 1844	Dec 1845	BU Mar 1879
Vulture	Pembroke Dyd	Sep 1841	21 Sep 1843	Jun 1844	BU 1866
Sampson	Woolwich Dyd	Apr 1845	1 Oct 1844	1846	BU Jul 1864
Centaur	Portsmouth Dyd	Dec 1844	6 Oct 1845	1847	BU 1864
Dragon	Pembroke Dyd	Jan 1844	17 Feb 1845	May 1847	BU Feb 1865
Avenger	Devonport Dyd	27 Aug 1844	5 Aug 1845	1846	20 Dec 1847, wrecked off N Africa

Other Second Class Frigates

Name	Builder	Laid Down	Launched	Completed	Fate
Birkenhead	Laird	Sep 1843	30 Dec 1845	1846	26 Feb 1852, wrecked off S Africa
Tiger	Chatham Dyd	Nov 1847	1 Dec 1849	Jun 1852	12 May 1854, burnt off Odessa
Magicienne	Pembroke Dyd	Aug 1847	2 Mar 1849	Nov 1852	BU Sep 1866
Valorous	Pembroke Dyd	Jan 1849	30 Apr 1851	Dec 1852	BU Feb 1891
Furious	Portsmouth Dyd	Jun 1848	26 Aug 1850	Nov 1852	BU 1885

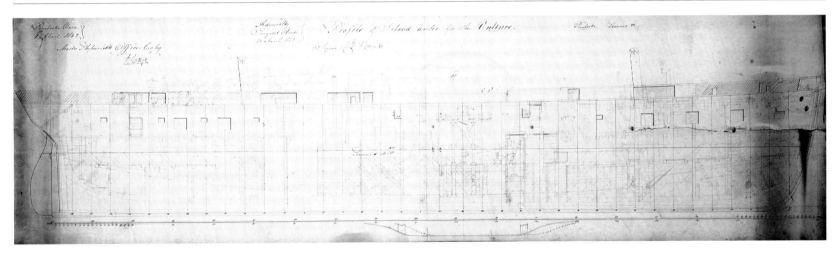

Table 21: *CYCLOPS* DERIVATIVES – SECOND CLASS FRIGATES
Machinery Particulars

Vulture, another second class frigate.

She differed from *Firebrand* in having two funnels, side by side. Note that she does not carry box boats but has smaller boats carried inboard of the paddle boxes.

Ship	Engines by	NHP	Type	No Cyl	ihp =	speed (trials)
Firebrand	Seaward	400	Direct	2	–	10.0
Gladiator	Miller	430	Direct	2	–	9.5
Vulture	Fairbairn	470	Direct	2	–	–
Sampson	Rennie	467	Direct	2	–	9.5
Centaur	Watt	540	Direct	2	–	10.5
Dragon	Fairbairn	560	Direct	2	–	11.5
Avenger	Seaward	650	Direct	2	–	9.5

Other Second Class Frigates

Ship	Engines by	NHP	Type	No Cyl	ihp =	speed (trials)
Birkenhead	Forrester	536	Direct	2	–	c11
Tiger	Penn	400	Oscillating	2	–	9-10
Magicienne	Penn	400	Oscillating	2	–	9-10
Valorous	Miller	400	Oscillating	2	–	9-10
Furious	Miller	400	Oscillating	2	–	–

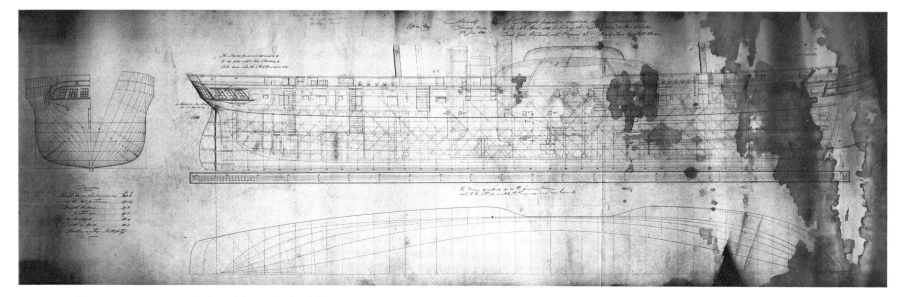

Sampson was a second class frigate derived from *Gorgon*.

The drawing shows her unusual arrangement of two funnels, close together and on the centreline abaft the paddles. She is shown with two masts but very soon she was altered to a barque and carried box boats. The vestigial gun ports are visible.

Table 22: *CYCLOPS* DERIVATIVES – SECOND CLASS FRIGATES
Armament

Where the date of a change in armament is not known, the outfits will be numbered 1, 2 etc. In earlier ships, all guns were carried on the upper deck. Later ships carried a main deck armament, shown as MD, as well as upper deck, UD. In general, data relate to the class, individual variations are indicated by the first two letters of the name.

Ship	Date	Armament
Firebrand	As comp	2x68pdr/95cwt pivot, 2x8in/65cwt,
Gladiator	1856	2x68pdr/95cwt pivot, 4x10in/84cwt
Vulture	1865 *(Gl)*	2x64pdr/71cwt RML pivot, 4x64pdr/71cwt RML
	Later *(Gl)*	2x110pdr/82cwt RBL pivot, 4x10in/84cwt
Sampson	1845	2x68pdr/95cwt pivot, 2x8in/65cwt
	1847-52	2x56pdr/87cwt pivot, 4x8in/65cwt, 2x24pdr/13cwt
	1856	2x68pdr/95cwt pivot, 4x10in/65cwt
Centaur	1845	2x10in/84cwt pivot, 4x68pdr/64cwt
Dragon	1856	2x68pdr/95cwt pivot, 4x10in/65cwt
	1861 *(Dr)*	2x110pdr/82cwt RBL pivot
Avenger	1	2x68pdr/95cwt pivot, 4x8in/65cwt, 4x32pdr/25cwt
	1847	2x8in/65cwt pivot, 8x32pdr/25cwt

Ship	Date	Armament
Other Second Class Frigates		
Birkenhead	As comp	2x10in/84cwt pivot, 4x68pdr/64cwt (possibly 6x68pdr)
	As troopship	4x32pdr/25cwt
Tiger	Design	MD 4x32pdr/56cwt UD 1x68pdr/95cwt pivot, 1x10in/84cwt, 4x32/42cwt
	1852	MD 10x32pdr/50cwt UD 2x10in/84cwt pivot, 4x32pdr/50cwt
Magicienne	As comp	MD 10x32pdr/50cwt UD 2x10in/84cwt pivot, 4x32pdr/50cwt
Valorous	1862	MD 8x32pdr/50cwt UD 2x110pdr/82cwt RBL pivot, 4x64pdr//1cwt RML
	1870 *(Va)*	MD 8x64pdr/58cwt UD 2x64pdr/71 RML pivot, 2x12pdr
	1875 *(Va)*	UD 2x64pdr/71 RML pivot, 4x64pdr/58cwt (No MD)
Furious	Design	MD 8x32pdr/56cwt UD 1x68pdr/95cwt pivot, 1x10in/84cwt, 4x32/42cwt
	1860	MD 10x32pdr/50cwt UD 2x10in/84cwt pivot, 4x32pdr/50cwt

Gladiator, built at Woolwich in 1845.

A second class frigate of the *Firebrand* class, derived from *Cyclops*. This photograph gives a good view of the box boats over the paddles.

Dragon. The last of the many *Gorgon/Cyclops* derivatives.

Slightly higher steam pressure and bigger engines increased her speed but the 11.5 knots recorded on trial is suspect, perhaps with the current. Steam pressure was increased to 8-10 lb/in^2 and she had 560 nominal horse power. The machinery framing is less elaborate than in the earlier ships though still massive.

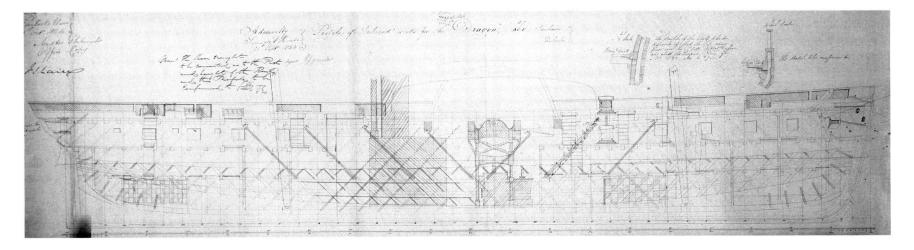

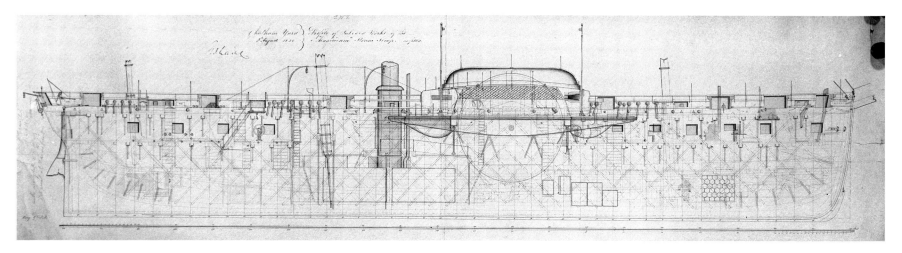

Magicienne was a second class frigate designed by Edye.   (a) Profile
She was a lengthened *Tiger* and had twin funnels, side by side.

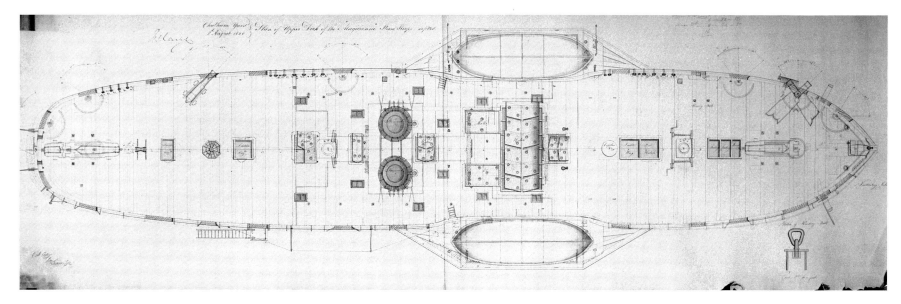

Tonnage

At this time the replacement of the traditional 'builders measurement' tonnage with 'new measurement' was under consideration, roughly the familiar 'gross tonnage' of merchant ships. It was noted that *Furious*, which was of 1282 tons on the old measurement, had a tonnage volume of 1365 on the new measurement from which would be deducted 561 tons for the machinery, giving 804 tons new measurement. A few documents do quote new measurement, usually with the old tonnage as well, but fortunately this never came into general use. There is enough confusion between tonnage, a measure of volume, and displacement, which is weight, without the intrusion of a third figure.

These twenty-nine ships, basically similar to *Gorgon*, reflect considerable credit on the Surveyor's Department and particularly on John Edye, and they form one of the largest groups of ships ever to have been built in peacetime.

(b) Upper deck plan. This shows her two funnels, side by side. It also shows the arcs of fire of her guns, including the alternative ports for the swivels fore and aft. The forward gun was just capable of axial fire and, from the after port, could fire abaft the beam. Moving to a new port would have been slow and hard work.

Furious was originally approved as *Sphynx* but much altered.

This ship was designed by Fincham in 1847 with six guns on the upper deck and eight or ten on the main deck.
(D K Brown collection)

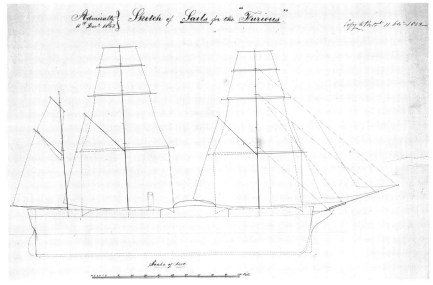

She has a main deck battery not shown on this sail plan. She was the last of 29 ships generally similar to *Gorgon*. The rig, shown for 1852, is typical of the later ships.

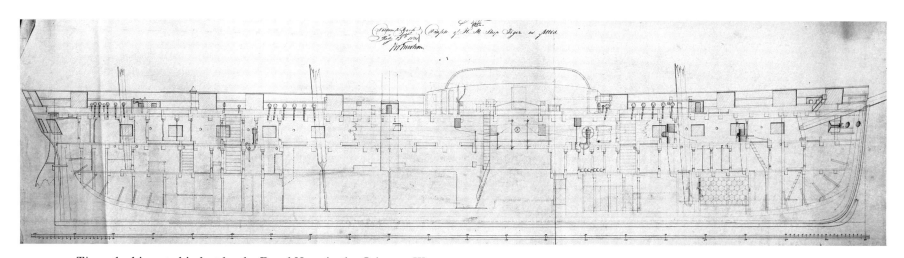

Tiger, the biggest ship lost by the Royal Navy in the Crimean War.

She was originally approved as a sloop similar to *Sphynx* but modified by Edye and became the first of the second class frigates to carry guns on the lower deck.

Model of *Tiger*

(Science Museum, London)

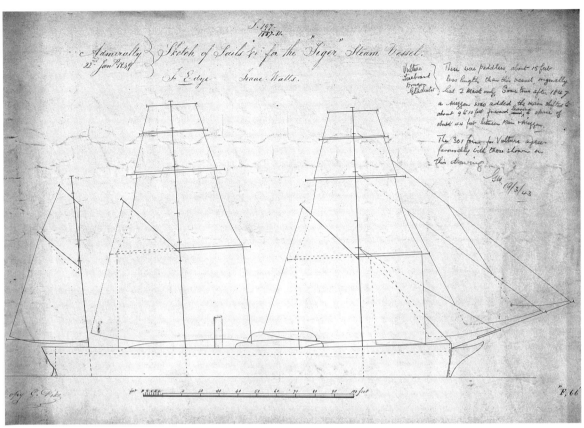

Tiger's sail plan, typical of later paddle frigates.

5. *Later Sloops and Second Class Frigates*

Trident, the first iron fighting ship to be ordered for the Royal Navy.

She was ordered from Ditchburn and Mare in 1843 as a sloop, third class.

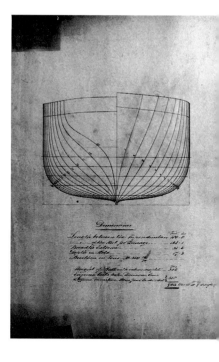

THE 1840s was a time of change, technically and also politically, with Symonds' loss of authority. In consequence, there were a considerable number of experimental vessels, not all of which were successful.

Janus was one such, designed by the Earl of Dundonald (Lord Cochrane) and double ended with vertical stem and stern. She had watertube boilers of Cochrane's own design and a rotary engine built in the USA. Her trials were abandoned in 1846 and in 1849, when she was reclassified as a tug, it was said that her engine was useless. It is possible that she was re-engined in 1854 when she was further reclassified as a gun vessel and, at least on paper, given a 10in 68-pounder pivot and four 32-pounders.

The Board invited tenders for an 'iron war steamer' in January 1843, the first iron warship ordered for the Royal Navy, though not the first to complete. Ditchburn & Mare's tender was accepted in April 1843 and she completed in 1846, as *Trident*; she was classified as a sloop, third class. Her initial armament consisted of two 10in pivots and two 32-pounder carronades. She had a Boulton & Watt oscillating engine of 350NHP which gave a speed of about 9½ knots.

In August 1845 the Surveyor was asked to design two mail steamers for service between the Ionian Isles, Malta and Greece. The next month a third ship was added and it was decided that they should have iron hulls. By the time the three *Antelopes* completed in 1847 they were classified as sloops, third class (later gun vessels) carrying a 68-pounder and two 32-pounder carronades.

Argus was approved in February 1847 as a sixth ship of the *Alecto* class but in August of that year it was decided to adopt new plans by Fincham. She was larger and heavier than *Alecto* and carried a heavier armament and was classified as a sloop, second class. On completion, her armament consisted of two 10in and four 32-pounder carronades. *Argus* was a barque with twin funnels abaft the paddles. *Buzzard* was approved in July 1847 to Edye's design and was generally similar to *Argus*. She cost £44,134.

Barracouta was a conventional second class sloop approved in February 1847. There was prolonged debate over her lines and only after discussion by the Committee of Reference were they approved in January 1848. She was the last sea-going paddle warship to be laid down for the Royal Navy.

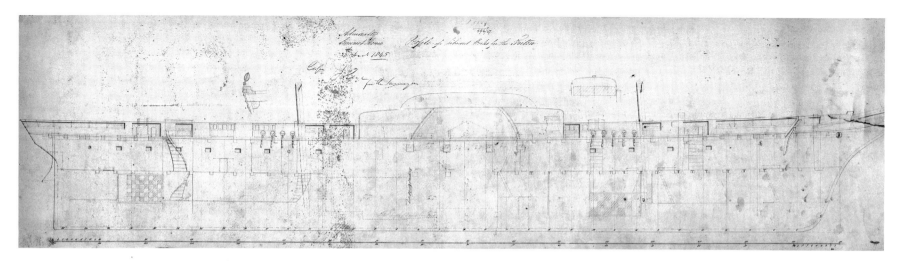

Triton, an early (1845) iron sloop of the *Antelope* class.

These three ships were quite small and were later reclassed as gun vessels.

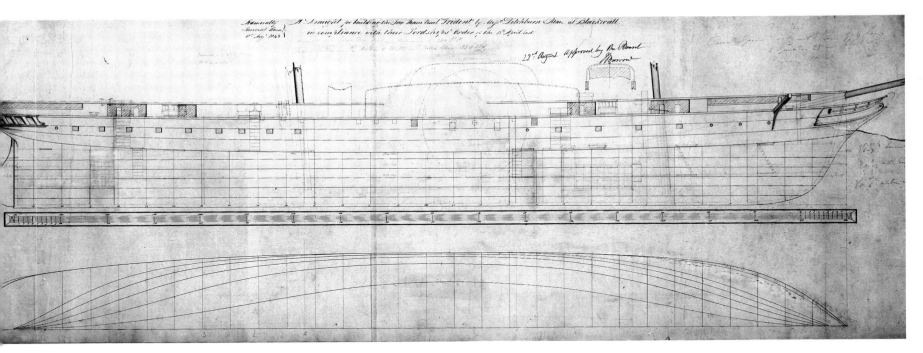

Later second class frigates

The *Birkenhead*, originally to have been *Vulcan* and renamed at launch on 22 June 1848, was one of the larger ships in the Admiralty's iron hulled programme of the early 1840s. An invitation to tender was prepared by Symonds and Sir Edward Parry of the Steam Department and sent to Lairds early in 1843. She was to use engines being built by Forrester for *Janus*. Lairds' tender was accepted in April 1843 and the lines were approved the following August. Forrester's tender of £26,132 was accepted in April 1844. Her armament of two 10in and four 68-pounders was on the upper deck.

Janus, an experimental sloop designed by the Earl of Dundonald (Cochrane).

She had experimental machinery with watertube boilers of Cochrane's design which was a complete failure. Her name indicates that she was double ended, steaming equally badly in either direction.

Table 23: LATER SLOOPS
Specification

Name	Tons bm	Disp't	Length ft in		Beam ft in		Depth ft in		Draught For'd ft in	Aft ft in
Trident	850	903	180	0	31	6	17	3	10 9mean	
Buzzard	980	1530	185	0	34	0	20	0	–	
Argus	981	1630	190	2½	32	5½	20	11	–	
Barracouta	1053	1676	190	2	35	0	20	5	–	
Janus	763	–	180	0	30	6	19	1	11 2mean	
Antelope	650	1055	170	8	28	3	18	6	c10 2	10 10
Oberon			172	4	28	0	18	0		
Triton			172	11½	28	2	18	0		

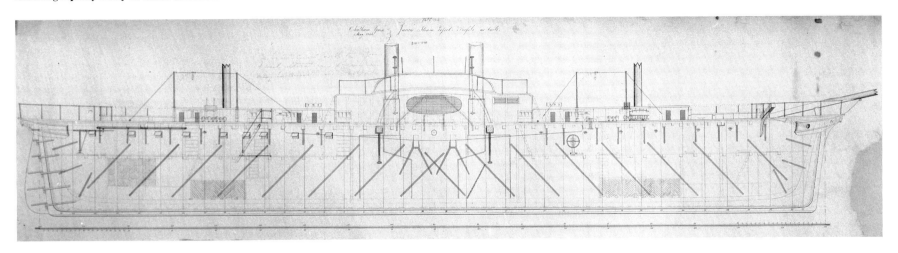

Table 24: LATER SLOOPS
Building Data

Name	Builder	Laid Down	Launched	Completed	Fate
Trident	Ditchburn & Mare	1845	16 Dec 1845	1846	BU Mar 1866
Buzzard	Pembroke Dyd	Oct 1847	24 March 1849	1851 Devonport	BU 1883
Argus	Portsmouth Dyd	Jun 1848	15 Dec 1849	1852 Woolwich	BU Oct 1881
Barracouta	Pembroke Dyd	May 1849	31 Mar 1851	Dec 1851 at East India Dock	BU 1881
Janus	Chatham Dyd	6 Sep 1843	6 Feb 1844	1846	BU Apr 1858
Antelope	Ditchburn & Mare	Dec 1845	25 Jul 1846	1847	BU Sep 1843
Oberon	Rennie	7 Jan 1846	2 Jan 1847	1847	BU Nov 1880
Triton	Wigram	3 Dec 1845	24 Oct 1846	1847	BU Feb 1872

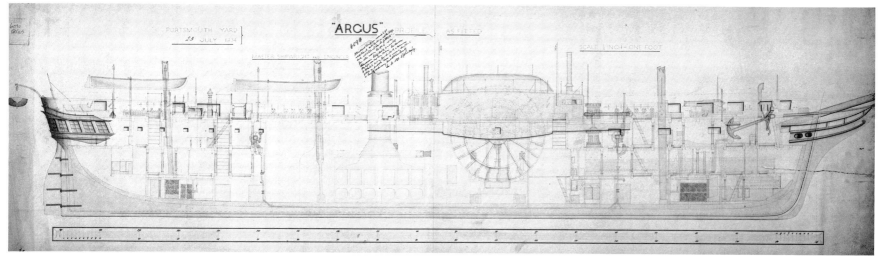

Argus was a second class sloop.

Originally approved as a sixth *Alecto*, she was redesigned by Fincham as a much larger ship.

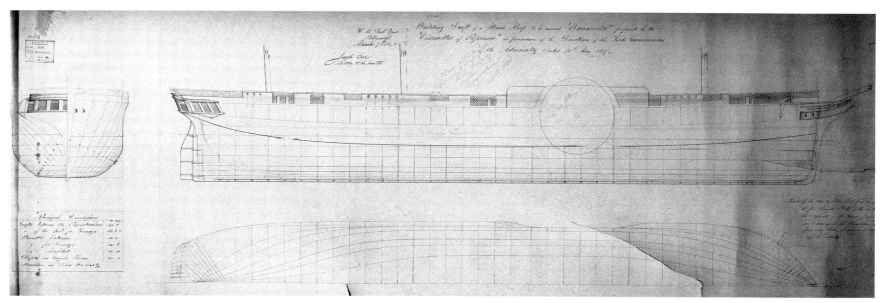

Barracouta was a conventional second class sloop approved in 1847.

She was caught up in the confusion which marked the decline of Symonds and her design was only approved by the Committee of Reference in January 1848.

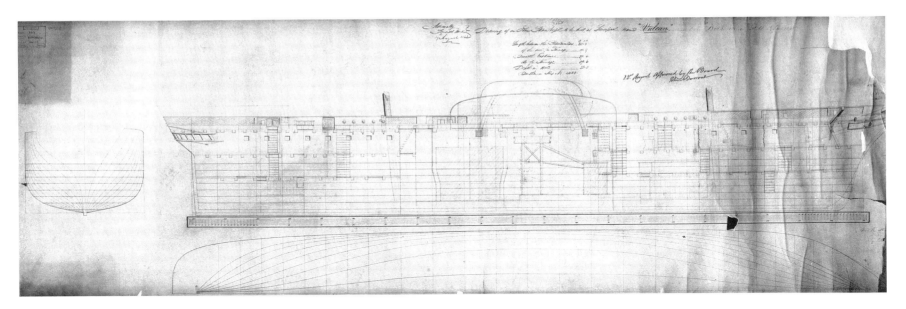

Birkenhead, the only iron paddle frigate.

She was originally to have been called *Vulcan*, as shown on this early plan. She became a troopship and her loss in 1852 was long remembered for the heroism displayed by the troops on board.

Table 26: LATER SLOOPS
Armament

Where the date of a change in armament is not known, the outfits will be numbered 1, 2 etc. All guns carried on the upper deck.

Ship	Date	Armament
Trident	1848	2x10in/84cwt pivot, 2x32pdr/25cwt
	2	2x10in/84cwt pivot, 4x32pdr/25cwt
	1856	1x32pdr/42cwt pivot, 2x32pdr/25cwt
	1862	1x40pdr/35cwt RBL pivot, 2x32pdr/25cwt
Buzzard	1851	2x10in/84cwt pivot, 4x32pdr/25cwt
	2	1x68pdr/95cwt pivot, 1x10in/84cwt pivot, 4x32pdr/25cwt
	3	1x110pdr/82cwt RBL pivot, 1x10in/84cwt pivot, 4x32pdr/25cwt
Argus	1852	2x10in/84cwt pivot, 4x32pdr/25cwt
	1856	1x68pdr/95cwt pivot, 1x10in/84cwt pivot, 4x32pdr/25cwt
	3	1x110pdr/82cwt RBL pivot, 1x10in/84cwt pivot, 4x32pdr/25cwt
Barracouta	1852	2x10in/84cwt pivot, 4x32pdr/25cwt
	1856	1x68pdr/95cwt pivot, 1x10in/84cwt pivot, 4x32pdr/42cwt
	1862	1x110pdr/82cwt RBL pivot, 1x10in/84cwt pivot, 4x32pdr/42cwt
Janus	Design	2x10in/84cwt pivot
	1855	1x68pdr/95cwt pivot, 1x10in/84cwt pivot, 4x32pdr/17cwt
Antelope	As comp	1x68pdr/95cwt pivot, 3x32pdr/17cwt
Oberon	1856	1x32pdr/45cwt pivot, 2x32pdr/25cwt
Triton		

Table 25: LATER SLOOPS
Machinery Particulars

Ship	Engines by	NHP	Type	No Cyls	ihp=	speed (trials)
Trident	(1) Maudslay	200	Side lever	2	–	–
	(2) Watt	350	Oscillating	2		9.5
Buzzard	Miller	300	Oscillating	2	853	10
Argus	Penn	300	Oscillating	2	764	10
Barracouta	Miller	300	Direct	2	881	10.5
Janus	(1) See text		Rotary			ca6
	(2) 1855?	220	Side lever	2 ex-*Sydenham*		
Antelope	Penn					
Oberon	Rennie	260	Oscillating	2		10.5-11
Triton	Miller					

She completed in 1846 as a brig but was soon converted into a barquentine. In 1847 she was used to tow the SS *Great Britain* off the shore of Dundrum Bay, and was put in reserve in 1848 as a result of the growing distrust of iron hulls. She was converted to a troopship in 1851 and during the conversion large holes were cut in the bulkheads to improve ventilation in the troop decks. These contributed to her rapid sinking after striking a rock in Danger Bay, South Africa, on 26 February 1852. It should be remembered that water-tight bulkheads were only introduced with iron hulls as timber ships worked too much for them to be practicable and their contribution to safety was not fully appreciated.

Avenger was approved early in 1844 to a new design by Edye with engines identical to those of *Penelope*. She was originally ordered from Deptford, but the work was transferred to Devonport where she was completed in 1856. She was wrecked off North Africa the following year but her armament of two pivots and eight broadside guns was changed even during this short life.

6. First Class Frigates

IN GENERAL, the first class frigates all had a heavy armament on the main deck as well as on the upper deck and this made them very powerful warships. There was a tendency for them to be faster than the second class vessels, with speeds of about 11 knots rather than the 9-10 typical of the smaller vessels.

Penelope was the earliest first class paddle frigate and unique as the only conversion to paddle from a sailing frigate. She was launched in 1820 as a sailing ship, sister to *Hebe*. Edye proposed that she be cut in half, lengthened by 65ft and given powerful engines, a scheme which was approved in March 1842. Seward & Capel's tender for 650NHP engines was accepted the following month. The machinery was quite advanced with four tubular boilers working at 8lb/sq in supplying steam to a two-cylinder direct acting engine which exhausted to a Hall's surface condenser. She was barque-rigged with iron wire for her standing rigging, possibly the first example of iron rigging in the Royal Navy,

Penelope was the only sailing frigate converted to paddle propulsion.

The conversion was planned by Edye who, unusually, was in error with his weight calculations and she floated deeper than intended. Edye hoped that there would be many more such conversions, but conversion to screw proved more economical.

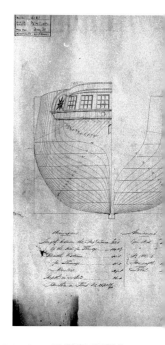

Penelope, from a contemporary watercolour. Because of Edye's incorrect weight estimates, she floated with a freeboard of only 4ft 11in to the gun sills.

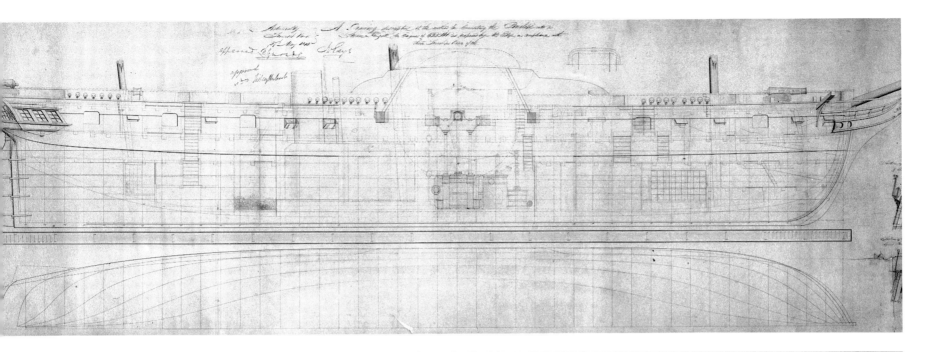

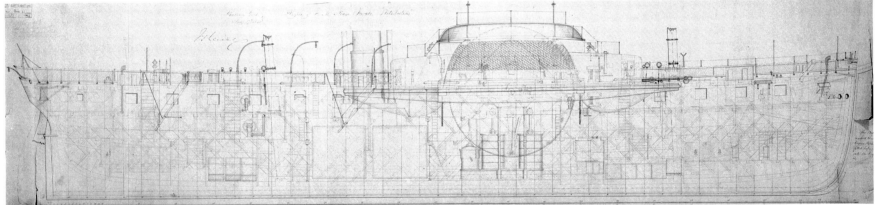

and proved successful. She had a single funnel between the main and mizzen masts.

Edye's first scheme was to install the engines in the original hull without lengthening and the extra length was intended mainly to give buoyancy, though the space must have been very welcome. His calculations, however, were incorrect and she completed much overweight and hence floated deeper. The bulwarks were raised, which helped to keep the deck dry and improved her appearance, but nothing could be done about the height of the main deck ports whose sills were only 4ft 11in above the water. Captain Jones said in 1843, 'In a heavy sea and a strong gale she steamed to windward without much pitching or straining and appeared dry, she answered [her helm] on all points, but under sail she was leewardly'.

Edye argued strongly for at least fifteen more such conversions but with smaller engines to save weight and cost, increase freeboard, and improve fuel economy. Nothing came of this proposal, probably because it was realised that converting to screw propulsion was more

Retribution, a first class frigate.

She was originally to have been called _Watt_ and, as completed, was not a success. The engines and boilers were both replaced and moved and the armament changed. After this, she was highly regarded.

effective and probably easier.

Retribution was designed by Edye and approved (as the _Dragon_) in March 1842. Her name was changed to _Watt_ in August when her NHP was also increased and, in 1844, she received her final name. Edye gave her a very flat floor and a brig rig but, in 1849, the main mast was moved forward by 6ft and a mizzen added.

Judging by the number of changes made in the first few years after completion in 1845, the design was not a success. The original flue boilers were replaced by tubular ones in 1847 and apparently moved 20ft aft to correct a 9in bow trim, after which she had twin funnels. Then, in 1849, her 800NHP Maudslay engines were replaced by 400NHP Penn oscillating engines driving 26ft cycloidal wheels in place

Terrible **was the largest and most powerful paddle frigate.**

This imposing view shows her as completed with four funnels. Two funnels and their boilers were removed about 1850.
(Science Museum, London)

Table 27: FIRST CLASS FRIGATES
Specification

Name	Tons bm	Disp't	Length ft in	Beam ft in	Depth ft in	Draught For'd ft in	Aft ft in
Penelope	1616	–	215 2	40 9	26 8	19 3	20 3½
Retribution	1641	–	220 0	40 6 (71 oa)	26 4	*c*18 mean	
Terrible	1850	3189	226 2	42 6	27 4	19 10 mean	
Odin	1310	–	208 0	37 2	24 3	18 0	18 11
Sidon	1316	–	211 0	37 0½	27 0	–	–
Leopard	1406	–	218 0	37 5½	25 0½	–	–

Table 28: FIRST CLASS FRIGATES
Building Data

Name	Builder	Laid Down	Launched	Completed	Fate
Penelope	Chatham Dyd	Nov 1827	13 Aug 1829	Converted Jun 1842–Apr 1843	BU Jul 1864
Retribution	Chatham Dyd	Aug 1842	2 Jul 1844	1845	BU Jul 1864
Terrible	Deptford Dyd	Nov 1843	6 Feb 1845	1845	BU Jul 1879
Odin	Deptford Dyd	19 Feb 1845	24 July 1846	1847	BU 1865
Sidon	Deptford Dyd	26 May 1845	26 May 1846	1846	BU Jul 1864
Leopard	Deptford Dyd	Aug 1846	5 Nov 1850	1851	BU Apr 1867

of the original 34ft common wheels. The armament was changed and the main deck battery fitted. She then had two pivots and eight heavy guns on the upper deck and twelve to eighteen 32-pounders on the main. After all these changes, *Retribution* seems to have been a satisfactory and well-liked ship.

In July 1842, Lang was invited to design a very large paddle frigate of 1640 tons bm, with 800NHP engines and, unusually, Lang was invited to 'state her armament'. Originally named *Simoon*, she became *Terrible* in December 1842. In October his plans were approved and building was ordered to start at Woolwich, but was transferred to Deptford in February 1843.

With a displacement of 3189 tons, she was the largest paddle warship and was very strongly built with her timber frames touching up as far

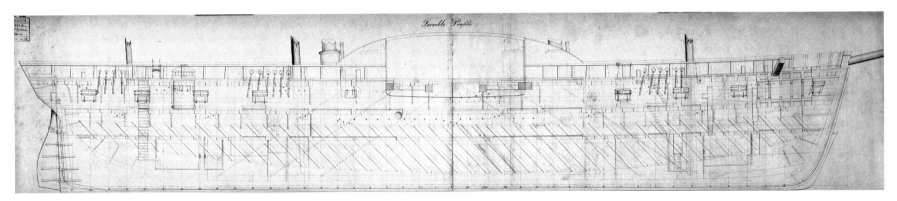

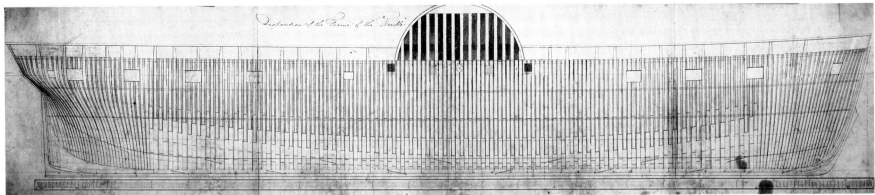

Terrible, **the most powerful paddle warship, built in 1843.**

She originally had four funnels, which can just be distinguished in the original print, but two funnels and their boilers were removed in about 1850 as they generated more steam than the engines could accept.

Her massive framing is well illustrated in the lower drawing. The lower frames were touching and she was watertight before the planking was added. This – and her imposing appearance – contributed to her long life in service.

as the waterline so that she was water-tight even before the planking was applied. She was an expensive ship at £94,650. Maudslay's tender of £40,259 for a 4-cylinder Siamese engine was accepted and there were four boilers working at 9 lb/sq in, each with its own funnel, giving her a most impressive profile. It would seem, however, that Maudslay had miscalculated as two boilers could supply all the steam that the engines could use. Two boilers and their funnels were removed about 1850.

She originally carried eight 56- and eight 68-pounders with four of each on the main and upper decks. *Terrible* remained in commission well into the 1860s and was well liked, not least for her appearance.

Odin was approved as a sloop of the *Sampson* class in August 1844 but in November it was decided to change to a new design by Fincham. The drawings were approved in February 1845 when it was decided that she should be built at Deptford where Fincham was Master Shipwright.

Fincham claimed that he intended to design a ship to carry a

Table 29: FIRST CLASS FRIGATES
Machinery Particulars

Ship	Engines by	NHP	Type	No Cyls	ihp =	speed (trials)
Penelope	Seaward	650	Direct	2	–	c11
Retribution	Maudslay	800	Direct	4	–	c10
	Penn (1849)	400	Oscillating	2	–	–
Terrible	Maudslay	800	Direct	4	2059	10.9
Odin	Fairbairn	560	Direct	2	–	10
Sidon	Seaward	560	Direct	2	–	10
Leopard	Seaward	560	Direct	2	–	11.2

broadside armament (he probably meant main deck), to reduce rolling and obtain a shallow draught.[20] The first aim was achieved for *Odin* carried six heavy guns on the upper deck and ten 32-pounders on the main. Her draught was about the same as other ships of her size and as ideas on rolling were fallacious until Froude's work in the 1860s, it is unlikely that this aim was achieved. She had two funnels on the centreline abaft the paddle boxes. Her coal was stowed in iron tanks which could be filled with water after the coal was used in order to keep the paddles at a constant immersion.

Sidon was designed by Captain Sir C Napier and approved in April 1845. He was allowed to use *Odin*'s lines but increased the depth by 2ft 9in. This increased the armament but reduced stability, and she had a reputation for being crank, though this gave her an easy, long period roll. Her captain noted that she was the 'best steam vessel I ever saw.

Table 30: FIRST CLASS FRIGATES
Armament

Where the date of a change in armament is not known, the outfits will be numbered 1, 2 etc.

Ship	Date	Armament
Penelope	1843	MD 2x42pdr/84cwt pivot, 10x42pdr/2cwt carronades UD 10x8in/65cwt
	1856	MD 10x8in/65cwt UD 2x10in/84cwt pivot, 4x8in/52cwt
	1862	MD 8x32pdr/50cwt UD 2x110pdr/82cwt RBL pivot, 2x64pdr/71cwt RML
Retribution	1843	UD 2x68pdr/112cwt, 4x8in/65cwt, 2x32pdr/40cwt (No MD)
	1845	UD 2x68pdr/112cwt, 6x8in/65cwt, 2x32pdr/34cwt, 2x24pdr/13cwt
	1850	MD 12x32pdr/50cwt UD 1x68pdr/112cwt, 1x8in/65cwt pivot, 8x8in/65cwt
	1856	MD 18x32pdr/50cwt UD 1x68pdr/95cwt pivot, 9x8in/65cwt
Terrible	1845	MD 4x56pdr/98cwt, 4x68pdr/95cwt, 3x12pdr/6cwt UD 4x56pdr/98cwt, 4x68pdr/95cwt
	1852	MD 4x56pdr/87cwt, 4x8in/65cwt UD 4x68pdr/95cwt, 4x8in/65cwt
	1862	MD 4x110pdr/82cwt RBL, 10x8in/65cwt UD 5x110pdr/82cwt RBL pivot, 2x68pdr/95cwt
Odin	1856	MD 10x32pdr/56cwt UD 2x68pdr/95cwt, 4x10pdr/84cwt
	1862	MD 6x32pdr/56cwt, 4x40pdr/35cwt RBL UD 5x110pdr/82cwt RBL, 1x68pdr/95cwt
Sidon	1	MD 14x8in/60cwt, 2x68pdr/88cwt pivot (fore & aft) QD 4x8in/52cwt on slides. Fo'c'sle 2x8in/ 52cwt on slides
Leopard	As comp	MD 12x32pdr/56cwt UD 2x68pdr/95cwt, 4x10in/84cwt
	1862	MD 8x32pdr/56cwt, 4x40pdr/35cwt RBL UD 5x110pdr/82cwt RBL, 1x68pdr/95cwt

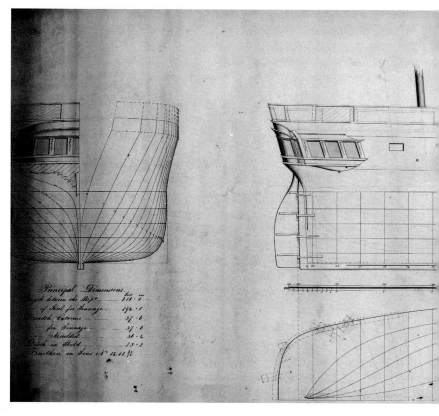

Leopard (1846), a first class frigate based on Fincham's *Odin* but with 10 feet extra length.

She carried six big guns on the upper deck and twelve 32 pounders on the main.

Sidon (below and opposite), designed by Sir Charles Napier.

She was based on Edye's *Odin* but the depth was increased, reducing her stability. Though she was a bit 'crank' she seems to have been satisfactory in service.

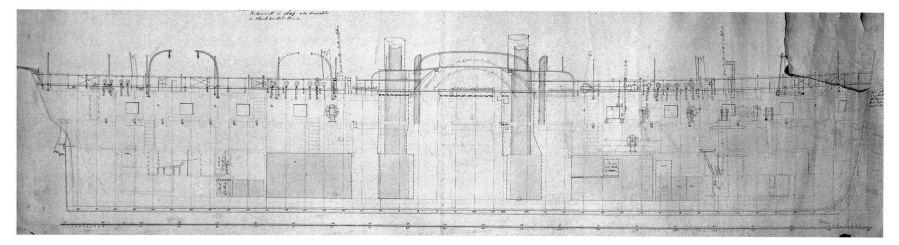

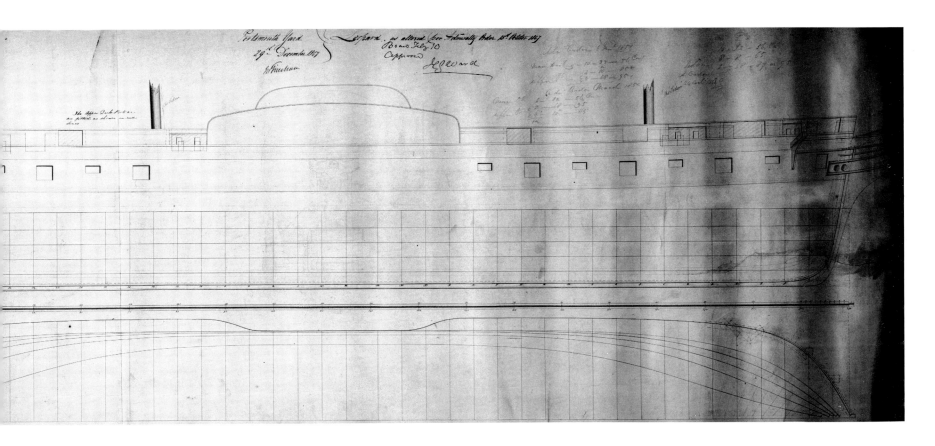

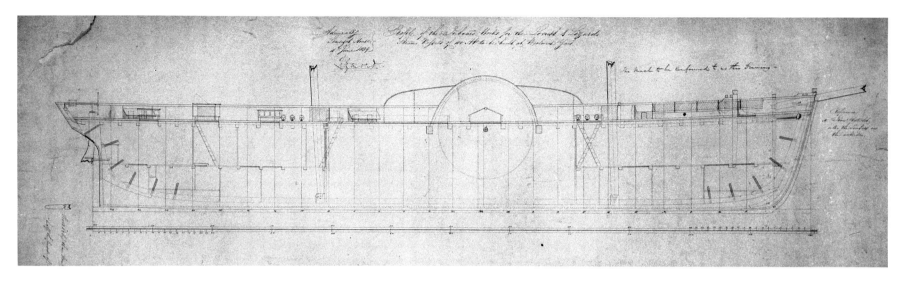

Lizard, an iron gunboat of 1845.

She was built and engined by Napier to lines by the Surveyor, Sir William Symonds.

Easy in a seaway, 500 men carried'. Her funnel was telescopic, perhaps the first such to be fitted.

Leopard was approved in March 1846 as a sister of *Odin* but in September her building was postponed, probably due to an overload of work at Deptford though it is recorded that her keel was laid in August 1846. She was re-approved in October 1846 but in January 1847 it was decided to wait until *Odin* had been tried. In February 1848 it was decided to lengthen her by 10ft. Other than this, she was generally similar to *Odin* but her boilers and their funnels were fore and aft of the engine room.

Though no more paddle frigates were built, designs were considered as late as 1852. These were all big ships, the largest being of 2540 tons/bm, (*cf Terrible*, 1850 bm) and would have carried thirty-six 68 pounders. There were to be no more paddle fighting ships, though a number of dispatch vessels, yachts and river gunboats were built.

7. *Miscellaneous Steam Vessels*

THERE WERE A number of other paddle vessels which did not, and still do not, fit into any logical classification or design sequence. Most were auxiliaries of little fighting value or technical interest; others were experimental vessels from which lessons could be learnt even if they were less than successful. Some were even designed to use an existing, spare engine. These ships are listed below, with brief notes, expanded if they are of interest.

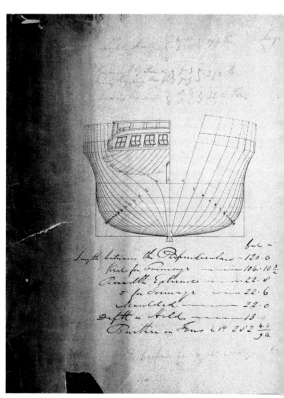

Locust, a small steamship whose classification was often altered.

She was approved in 1837 and, like the earlier generation of steam vessels, was an odd job ship – tug, tender or gunboat as required.

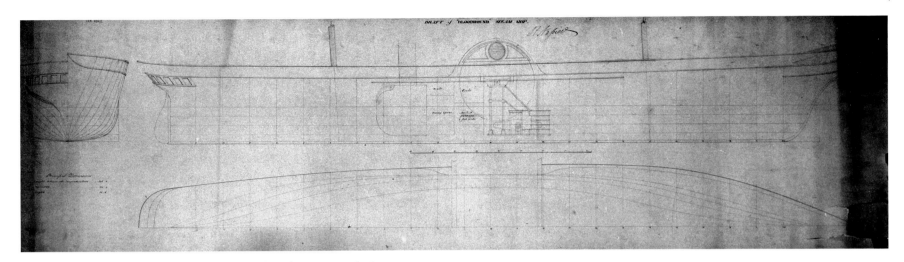

Falcon was a sailing brig and was launched in 1820 and given an experimental engine in 1833 which was removed the following year.

Lizard and *Locust* were approved in 1837 'as *Comet*' but said to be designed by Symonds. They were first classified as tenders, then gun vessels, 2nd class, and finally as tugs. *Lizard* had three-pounders and *Locust* one 32-pounder and two 12-pounder howitzers. *Locust* cost £13,800.

Porcupine. She was said to have been designed in 1843 by 'the Board of Admiralty' to use a spare Maudslay engine (ex-*Hermes*) and was used extensively as a gun vessel, survey vessel and tender. She had two funnels, fore and aft of the machinery, and her armament usually consisted of three 32-pounders.

Bloodhound was one of a group of six iron gunboats of 1845.

There were a number of differences in lines and power between the vessels: in the case of *Bloodhound* the lines and the 150NHP machinery were by the builders, Napier.

Spitfire was the second steamship of that name. She was designed in 1844 around a surplus engine (ex-*Firebrand*) and was generally similar to *Porcupine*. (The source of the engines for these two ships is that given in an Admiralty letter (ADM 12) of 11 November 1843 but there is a possibility that the final allocation was reversed.)

Though Post Office packets are dealt with in the Appendix, brief mention should be made here of *Dover*, the first iron-hulled ship ordered

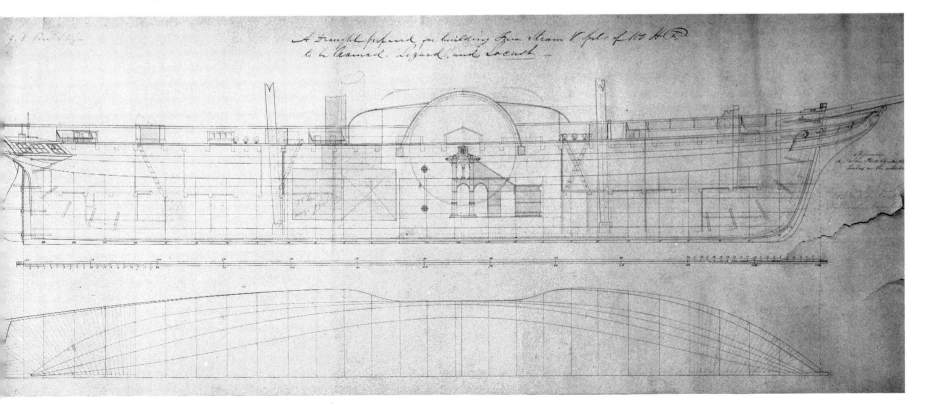

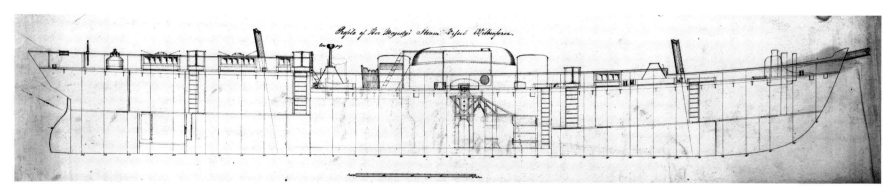

Wilberforce (above) and *Albert* (below) were built in 1840 for use on the river Niger.
They were designed by Lairds and were very early iron ships. The light weight of their iron hulls contributed to their shallow draught.

by the Admiralty, on 4 February 1839, shortly after Sir George Airy's work on the correction of compasses made it possible for iron ships to go out of sight of land. Though the Admiralty's experiment was cautious, they took considerable care, over a number of years, to compare the running costs of *Dover* with wooden ships on the same service. There was little evidence of any difference.

The East India Company ordered the gun vessel *Nemesis* at about the same time as *Dover* and the former's performance in the China War influenced the Board to build a considerable number of iron warships.

The *Jackall* class consisted of six gun boats and *Jackall* herself was the first iron warship to complete for the Royal Navy in March 1845. Tenders for the class were invited in January 1844 but the programme was changed several times. There were two shipbuilders, Napier and

Ditchburn & Mare, and it was decided that the first ship from each builder, *Lizard* and *Torch* respectively, should use Symonds' lines with 150NHP side lever engines. The next pair, *Jackall* and *Harpy*, were to have the same lines but engines of the highest power available which could be fitted. The last two, *Bloodhound* and *Myrmidon*, were to have lines by the builders and 150NHP engines. Trial reports are dubious but suggest speeds of between 9 and 11 knots.

They seem to have been very useful ships and remained on active service for many years. *Lizard* and *Harpy* took part in the Parana River

Bee was an instructional tender for the Naval College.

She had both paddles and a propeller but this beautiful drawing shows an early machinery installation for paddles only.

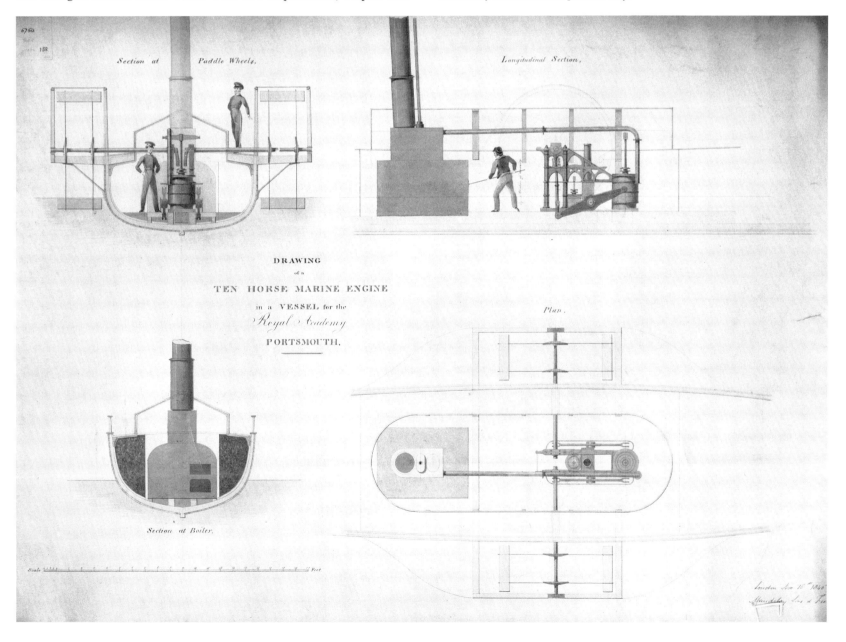

Section at Paddle Wheels.

Longitudinal Section.

DRAWING
of a
TEN HORSE MARINE ENGINE
in a VESSEL for the
Royal Academy
PORTSMOUTH.

Plan.

Section at Boiler.

Recruit of 1851, built by Scott Russell.

Originally laid down for the Prussian navy, she was purchased for the RN at the outbreak of the Crimean War. She was much overweight, and her armament was later reduced.

campaign in 1846 following which arguments over the nature of the damage they received were important in the debate over the future of the iron fleet. *Harpy*, c1892, had the unique distinction of being sunk during trials of the Zalinski dynamite air gun.

Grappler was designed and built by Fairbairn in 1844, again using a spare engine. She had an iron hull and, most unusually, it suffered from severe early corrosion and so she was disposed of in 1850. The hull cost £15,524, the machinery £17,790, and she was sold for £550. Though the smaller iron warships remained in commission when the iron frigates were sold or relegated to auxiliary duties, no more iron ships were built until 1855.

Recruit (ex-*Salamander*) and *Weser* (ex-*Nix*) were being built by Scott Russell for Prussia when the Crimean War broke out and they were bought into the Royal Navy, Prussia receiving the sailing frigate *Thetis* in exchange. (A third sister was delivered to Prussia.) They had iron hulls with 8in teak backing. They were double ended and, on trial, were able to steam at 11½ knots in either direction though their performance in service was disappointing since they completed much overweight. They carried four 8in, fore and aft of the paddle boxes, either side, and two 32-pounder carronades.

Bann and *Brune* were ordered from Scott Russell in 1855 for £11,400 each and were smaller but similar in style to the Prussian ships. They mounted two 8in and two 32-pounder carronades.

Albert, *Wilberforce* and *Soudan* were three differing iron ships built by Lairds in 1840 for use on the Niger. They took advantage of the light weight of an iron hull to achieve a shallow draught.

Bee was approved in September 1840 as an instructional tender for the Naval College. Her engine could work either a screw, the paddles or both together – even in opposition. She was the first ship with a screw to be ordered for the Royal Navy.

Helicon built at Portsmouth between 1861 and 1865.

Helicon and her sister ships were rated as despatch vessels, and used as yachts for Commanders-in-Chief. Her bow had the appearance, but not the strength, of a ram, giving her an extra knot over her sisters.

Table 31: MISCELLANEOUS PADDLE WARSHIPS
Specification

Name	Tons bm	Disp't	Length ft in	Beam ft in	Depth ft in	Draught For'd ft in	Aft ft in
Falcon	237	–	90 1	24 9	11 0	–	
Lizard and *Locust*	283	470	120 0	22 8½	13 0	8 6	
Porcupine	382	490	141 0	24 1½	13 6	–	
Spitfire	432	500	147 2	25 1	14 6	–	
Harpy and *Torch*	344	} 505	141 0	22 6	13 7	–	
Myrmidon	374						
Jackall and *Lizard*	340	} 505	142 7	22 6	12 10	–	
Bloodhound	378						
Grappler	557	–	165 2	26 7	16 7	–	
Recruit and *Weser*	540	468	178 0	26 0	10 6	7 0	
Bann and *Brune*	267	291	140 0	20 0	8 6	4 0	
Albert and *Wilberforce*	457	340	136 0	27 0	10 6	6 0 (level)	
Soudan	249	253	110 0	22 0	8 2	4 6	
Bee	42	–	60 0	12 4	6 0	3 5	
Rocket	70	–	90 0	12 8	7 4	–	
Ruby	73	–	90 0	12 9½	7 1	–	
Enchantress	835						
Psyche	835						
Helicon	837	*c*985	220 0	28 2	14 6	6 2	6 2 light
Salamis	835					10 2	10 6 load
Lively	842						
Vigilant	835						
Pioneer	295	–	153 0 (oa)	20 0	8 6	2 6	

Helicon and her sisters were designed as despatch vessels and yachts for Commanders-in-Chief. The first two, *Enchantress* and *Psyche*, were designed by Isaac Watts, *c*1860, and had a conventional knee bow. *Helicon* was modified by Edward Reed and given a long protruding plough bow with the shape but not the strength of a ram. *Salamis* was similar but with a vertical bow. *Helicon* was over a knot faster than the others as her stem was the equivalent of increasing waterline length and may also have acted as a bulbous bow. (The trials data is not altogether convincing; see Table 33.) *Lively* and *Vigilant* were similar to *Helicon*. They were very handsome ships, each with two bell mouthed funnels and a light rig. Each cost between £41,000 and £45,000.

Pioneer was a most unusual British paddle warship, built in Sydney at the yard of the Australian Steam Navigation Company in 1863 for service in the Maori War. She was an iron stern wheeler with two turrets based on Coles' design, each 12ft diameter and 8ft high. It was intended that each turret should mount a single 21-pounder but it seems that only 12-pounders were installed. She is said to have had 3.8in armour[21] though this seems unlikely. *Pioneer* arrived in New Zealand on 3 October 1863 and went into action shortly afterwards. The turrets were not successful and were removed in December 1863 to reduce her draught, and one of these is preserved in Hamilton. After the war she was used as a transport until 23 December 1866, when she broke from her moorings at Port Waikato, drifted ashore and broke up.

There were a few more paddle steamers in the Royal Navy, built as yachts or river gunboats, but the day of the paddle fighting ship was over. In World War I a number of paddle minesweepers were built, some of which survived to join later commercial vessels as auxiliary anti-aircraft ships, defending the Thames Estuary during World War II. After that war a few paddle tugs were built and, in about 1960, the idea of paddle minesweepers was re-examined but, though not without attractions, was abandoned in favour of the more conventional *Hunt* class.

Table 32: MISCELLANEOUS PADDLE WARSHIPS
Building Data

Name	Builder	Laid Down	Launched	Completed	Fate
Falcon	Pembroke Dyd	Jun 1833	1820	–	BU 1837
Lizard	Woolwich Dyd	Jul 1839	7 Jan 1840	1840	24 Jul 1843, collided off Spain
Locust	Woolwich Dyd	Jul 1839	18 Apr 1840	1840	BU 1894
Porcupine	Deptford Dyd	Jan 1844	17 Jun 1844	1844	BU Jul 1883
Spitfire	Deptford Dyd	Jun 1844	26 Mar 1845	1845	BU 1888
Lizard	Napier Glasgow	–	28 Nov 1844	1845	BU 1869
Jackall	Napier Glasgow	–	28 Oct 1844	1845	BU 1887
Bloodhound	Napier Glasgow	1844	9 Jan 1845	1845	BU 1866
Harpy	Ditchburn &	–	Mar 1845	–	BU 1909
Myrmidon	Mare,	1844	Feb 1845	1846	Sold 1 Dec 1858
Torch	Blackwall	–	Feb 1845	–	Sold May 1856
Grappler	Fairbairn	1844	30 Dec 1845	1846	BU Feb 1850
Recruit	Scott Russell,	–	1851	Purchased 12 Jan 1855	Sold Oct 1869
Weser	Millwall	–	1851	Purchased 12 Jan 1855	Sold 29 Oct 1873
Bann	Scott Russell,	2 Jan 1856	5 Jul 1856	–	Sold 18 Feb 1873
Brune	Millwall	2 Jan 1856	30 Aug 1856	–	Sold 19 May 1863
Albert	Laird	Aug 1840	Sep 1840	–	BU 1850
Wilberforce	Laird	Aug 1840	1840	–	BU 1850
Soudan	Laird	Aug 1840	1840	–	1844, wrecked, Africa
Bee	Chatham Dyd	14 Jun 1841	28 Feb 1842	–	BU Sep 1874
Rocket	Fairbairn	–	1842	–	BU 1850
Ruby	Acreman	–	1842	–	BU 1846
Enchantress	Pembroke Dyd	8 Feb 1861	2 Feb 1862	1863	BU 1869
Psyche	Pembroke Dyd	8 Jan 1861	29 Mar 1862	1862	15 Dec 1870, wrecked, Catania
Helicon	Portsmouth Dyd	30 May 1861	31 Jan 1865	1886	BU Jul 1905
Salamis	Chatham Dyd	10 Aug 1861	19 May 1863	1865	BU Oct 1873
Lively	Sheerness Dyd	1870	10 Dec 1870	1871	7 Jun 1883, wrecked, Stornaway
Vigilant	Devonport Dyd	1870	17 Feb 1871	1873	Sold 1886, China
Pioneer	Sydney	1863	16 Jul 1863	Oct 1863	23 Dec 1866, grounded

Pioneer was an unusual stern-wheel gunboat.

Built in Sydney in 1863 for use during the New Zealand Wars, she is shown as completed (right), and stripped down as a transport at Port Waikato, where she ended her days (left).
(Auckland Public Library)

Table 33: MISCELLANEOUS PADDLE WARSHIPS
Machinery Particulars

Ship	Engines by	NHP	Type	No Cyl	ihp =	speed (Trials)
Falcon	Maudslay	100	Side lever	2	–	–
Lizard and *Locust*	Watt	100	Side lever	2	157	–
Porcupine	Maudslay	132	Side lever	2	285	–
Spitfire	Maudslay	120	Side lever	2	–	–
Harpy	Penn	200	Side lever	2	520	9.0
Torch	Seaward	150	Side lever	2	–	11
Myrmidon	Watt	150	Side lever	2	–	10
Jackall, Lizard and *Bloodhound*	Napier	150	Side lever	2	455	–
Grappler	Maudslay	200	Siamese	4	–	–
Recruit and *Weser*	Russell	160	Oscillating	2	754	11½
Bann and *Brune*	Russell	80	Oscillating	2	364	10½
Albert and *Wilberforce*	Forrester	70	–	–	–	–
Soudan	–	35				
Bee	Maudslay	10	–	1	–	–
Rocket	Fairbairn	20	–	2		
Ruby	–	20	–	2		
Enchantress	Penn	250	Oscillating	2	1318	14
Psyche	Penn	250	Oscillating	2	1440	–
Helicon	Miller	250	Oscillating	2	1610	14.5
Salamis	Miller	250	Oscillating	2	1440	13.25
Lively	Penn	250		2	1750	–
Vigilant	Watt	250	Oscillating		1815	13.2
Pioneer	–	60		2		c9

Table 34: MISCELLANEOUS PADDLE WARSHIPS
Armament

Ship	Date	Armament
Falcon	1	8x18pdr, 2x6pdr. Later 6 small guns.
Lizard	1	1x18pdr/15cwt, 2x18pdr/10cwt
Locust	1	1x32pdr/34cwt
	2 (both)	1x32pdr/34 pivot, 2x12pdr/10cwt howitzers
Porcupine	1	1x32pdr/25cwt, 2x32pdr/17cwt
	2	1x18pdr/20cwt pivot, 2x18pdr/10cwt carronades
Spitfire	1	1x32pdr/25cwt, 2x32pdr/17cwt
Harpy *Torch* *Myrmidon* *Jackall*	Design (all)	1x18pdr/22cwt pivot, 2x24pdr/13cwt carronades, later 1 to 4 carronades depending on duty
Lizard	1846	1x32pdr/56cwt or 1x8in/65cwt pivot
Bloodhound	1856	1x18pdr pivot, 1x12pdr/10cwt howitzer
Grappler	1	1x32pdr/25cwt, 2x32pdr/17cwt
Recruit and *Weser*	Design 2	4x8in/65cwt, 2x32pdr/25cwt / 4x32pdr/56cwt, 2-12pdr/10cwt howitzers
Bann and *Brune*	Design 1860	2x8in/65cwt, 2x32pdr/25cwt / 2x32pdr/25cwt
Albert and *Wilberforce*	Design	3x12pdr, 4x1pdr
Soudan	Design	1x12pdr howitzer
Bee		
Enchantress, Psyche, Helicon, Salamis, Lively and *Vigilant*	Design	2x20pdr
Pioneer	Design 2	2x21pdr / 2x12pdr in two turrets

Classification of paddle warships

Early official references speak of 'Steam Engine Vessels' but in December 1827 *Echo, Lightning* and *Meteor* were rated as 'Steam Vessels' and by October 1828 all steam vessels were so listed. In December 1834 steam vessels were split into two classes:

SV   145ft long, 100NHP

SV   155ft long, 140NHP

A much more detailed classification was made on 21 January 1837 (Admiralty Order 21.1.37) when steam vessels were divided into five classes. As far as possible, dimensions of masts and spars were standardised within each group as were armament and stores outfit (boatswain, carpenter, engineers).

SV1   *Gorgon, Cyclops*

SV2   *Medea, Salamander, Phoenix, Rhadamanthus, Dee, Messenger*

SV3   *Hermes, Volcano, Megaera, Spitfire, Firefly, Firebrand, Flamer*

SV4   *Blazer, Tartarus, Columbia, Pluto*

SV5   *Lightning, Meteor, Confiance, Echo, Alban, Carron, African, Comet*

During 1839-41 some Post Office packets were added to the following classes.

SV4   *Kite, Avon, Gleaner, Lucifer, Shearwater*

SV5   *Boxer, Fearless, Monkey*

In 1844 Admiral Sir George Cockburn (Senior Naval Lord) and John Edye devised a new classification which was intended to standardise armament, though available data suggests that this was only partially implemented. (See class tables)

From the previous list, the following were promoted:

SV5 to GV1   *Spitfire, Porcupine*

Others were demoted:

SV1 to Sloop 1   *Gorgon*

SV2 to Sloop 2   *Medea, Phoenix, Salamander, Hydra, Hecla, Hecate, Dee* and *Rhadamanthus* to transports

SV3 to GV 1   *Firefly, Flamer, Sydenham, Firebrand* to yacht

SV3 to GV 2   *Kite, Avon, Gleaner, Shearwater*

SV5 to tender   *African, Carron, Comet, Confiance, Echo, Meteor, Fearless, Monkey, Dwarf, Alban* to transport

On 10 November 1847, John Edye proposed a new classification for masts and spars, taking into account the fitting of a mizzen to several classes. This was approved by the Board on 11 November.

Class

1   None

2   *Terrible, Retribution, Penelope, Avenger;* also *Sidon, Odin, Leopard* (masted by constructor)

3   *Centaur, Sampson, Dragon, Magicienne, Valorous, Tiger, Furious, Gladiator, Firebrand, Vulture, Cyclops, Gorgon*

4   *Vixen, Bulldog, Inflexible, Scourge, Fury, Sphynx, Rosamond, Devastation, Driver, Geyser, Styx, Growler, Virago, Cormorant, Spiteful, Stromboli, Vesuvius, Buzzard, Argus*

5   *Hydra, Hecla, Hecate, Salamander, Medea, Hermes*

6   *Ardent, Alecto, Polyphemus, Trident, Volcano, Acheron*

7   *Blazer, Tartarus, Firefly, Grappler, Flamer*
iron packets: *Oberon, Triton, Antelope*

8   *Alban, Spiteful*

1844 Classification	New Classification		Rank of CO	No. of Lts	Crew	Armament
SV1	Steam frigate	1st class	Capt	4	200	2x8in/112, 4x8in/65, 4-32pdr/25
SV1	Steam frigate	2nd class	Capt	3	175	2x8in/112, 4x8in/65, 2-24pdr
SV2	Steam sloop	1st class	Cdr	3	145	2x42pdr/84, 2x68pdr/64, 2-42pdr/24
SV3	Steam sloop	2nd class	Cdr	2	100	2x10in/84, 2x32pdr/25
SV3	Steam sloop	3rd class	Cdr	2	100	1x68pdr/65, 4x32pdr/17
SV4	Steam gun vessel	1st class	Lt	–	60	1x32pdr/26, 2x32pdr/17
SV5	Steam gun vessel	2nd class	Lt	–	60	1x18pdr/22, 2x18pdr/15

Valorous, a second class frigate, completed at Pembroke Dockyard in 1852.

Valorous was typical of the later, bigger ships and carried a very powerful armament. She was the last paddle frigate to remain in service in the Royal Navy.
(D K Brown collection)

Part II: The Ships

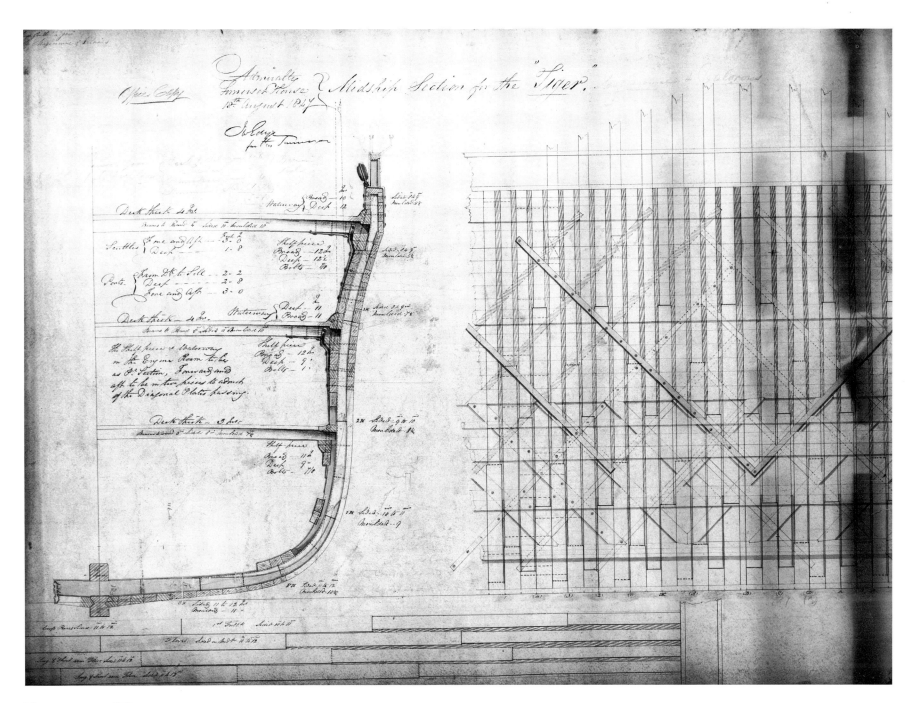

The structure of *Tiger*.

A clear illustration of the massive strength of paddle warships. Apart from the size, it is little different from the earlier *Hermes*. There was no need to change as the earlier ships had proved very strong in service.

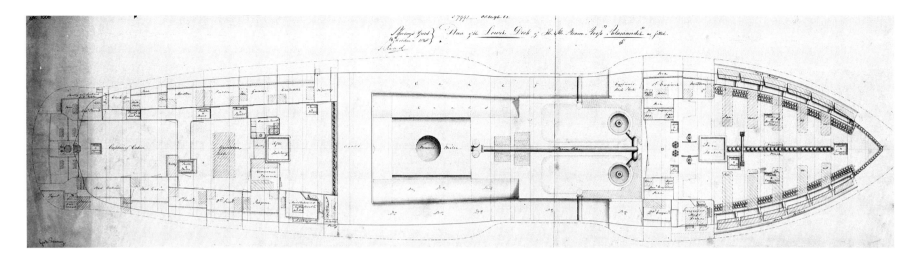

8. Design and Structure

THE MAJORITY OF paddle warships followed a similar pattern in design, which came about simply as a matter of practical considerations. The machinery, for example, was big and heavy so that it had to be placed amidships, which was also the most suitable position for the paddle wheels. From the start, it was realised that the slow running engines would cause vibration and the structure would have to be unusually strong and rigid, and this gave most of them a very long life. Another reason for similarity in design was that all the early ships, and a number of later ones, were designed by Oliver Lang and most of the rest by John Edye.

The middle third of the ship's interior was occupied by the machinery whilst, outside, the paddles and, in later ships, their sponsons obstructed about the same length, making it difficult to arrange a broadside battery. In almost all these ships the guns were in the open on the upper deck. There was a heavy pivot gun at each end, which required extra buoyancy and consequently fairly full lines underwater. These pivots could fire on the broadside and there would also be a few guns mounted on either broadside, usually just fore and aft of the paddles.

With the guns on the upper deck, the main deck, fore and aft of the machinery, was available for accommodation. Right forward, there was a big, open mess deck, where some 100 to 150 men lived in much the same way as in Nelson's ships. In this forward deck area there were also cabins for the warrant officers, master, gunner etc and for the engineers (usually two) and a small mess for midshipmen. The after part of the main deck contained spacious quarters for the commanding officer, the wardroom and small cabins for about three other officers. There was also a big chart room. Crew numbers varied from time to time but the figures given in the 1837 classification are reasonably typical.

The early ships had circular paddle boxes without sponsons. Sponsons, once introduced, provided convenient sites for many services, particularly those which needed fresh air or whose smell was

Salamander's lower deck, showing the accommodation.

The officers live aft in comparative luxury whilst the ratings have much the same area forward. About 100 men lived in the forecastle, which is some 1100 square feet in area. Poor ventilation and the heat from the boilers must have made conditions very unpleasant in the tropical waters where many such ships served.

better kept out of the hull. Such spaces included the galley with its beef screen, and the heads. Some ships also had a small arms store, handy for use in action.

There do not seem to be any accounts by ratings of life on board these early steam ships but it cannot have been comfortable. There was little space and the combination of inadequate ventilation and the heat from the boilers must have made them very hot in the tropics where many ships served.

The great majority of paddle warships were of wooden construction with substantial frames and, usually, a double layer of planking, 4in to 7in thick, depending on the size of the vessel. Both Lang and Edye

Tartarus. **Even in this early ship the heavy scantlings of the structure are noteworthy.**

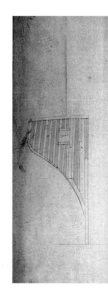

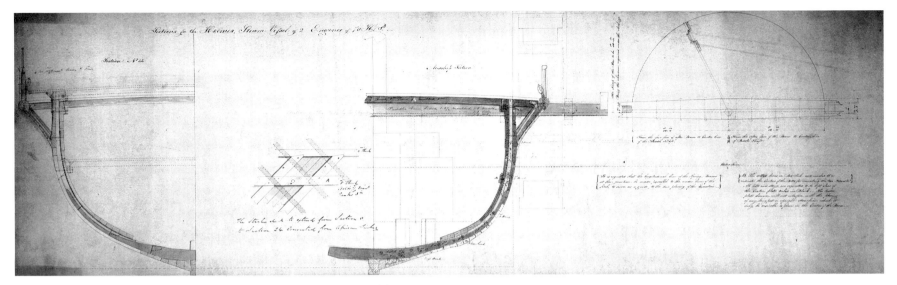

adapted Seppings' frigate system of diagonal framing with a considerable number of diagonal iron straps which helped considerably in preventing the ships from 'working', leading to early rotting and permanent bending of the hull, or 'breaking the sheer'.[22] Most of the wood-hulled ships were built in the Royal Dockyards.

The Admiralty's work on compass correction, led by Sir George Airy and published in 1838, made it possible for the first time for an iron ship to go out of sight of land. The Admiralty ordered a small iron packet, the *Dover*, for comparative evaluation of iron hulls. Following the success of the East India Company's gun vessel, *Nemesis*, in the First China War and of the Mexican paddle frigate *Guadaloupe* in the war against Texas (with an RN Captain), the Board embarked on an ambitious programme of iron warships of which a frigate and a number of smaller ships were paddle driven.

An iron hull weighs only about 80 per cent of a wooden one and will have some 20 per cent more usable internal space because the frames

Hermes, structural details.

The two sections show the strength of these early paddlers, whilst the insert between them shows the diagonal straps over an inner layer of diagonal planking, increasing her strength in shear. The right hand section shows one of the massive beams which supported the paddles, also shown in elevation in the right hand drawing.

are smaller. An iron hull, riveted so that the parts cannot 'work' is much more rigid and better able to resist vibration. The reasons, both political and technical, for the failure of this programme are complex but the technical aspect was the poor performance of wrought iron against gunshot, particularly in cooler weather when iron becomes brittle.[23]

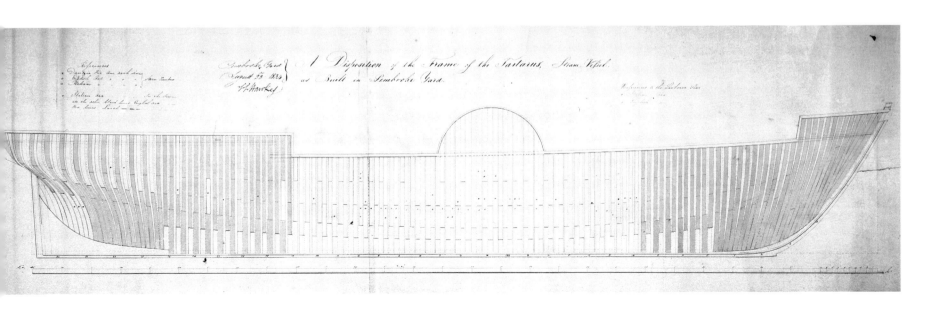

The side lever engines of *Dee*, 1825

These engines may be seen as a beam engine in which the beam is cut longitudinally, and dropped down beside the engine, lowering the centre of gravity and reducing the overall height. Note the gothic engine framing, introduced by Isambard Brunel.
(Science Museum, London)

9. *Machinery and Paddle Wheels*

THE FIRST STEAM ship ordered for the Navy, *Congo*, had a traditional beam engine but this configuration was quite unsuitable for ships as the heavy beam, high up, reduced stability and was vulnerable to gunfire. By the time that the Admiralty started building steam ships regularly, the side lever engine had been devised. It can be thought of as an engine in which the beam had been cut in half along its length and each half dropped down to lie either side of the lower part of the cylinders, with a linkage to connect it to the piston. Such engines were universal until the late 1830s and, since the loads were well balanced, vibration was not unduly severe.

Two-cylinder engines were used as there was always the risk that a single piston might stick at 'dead centre' making it impossible to start.

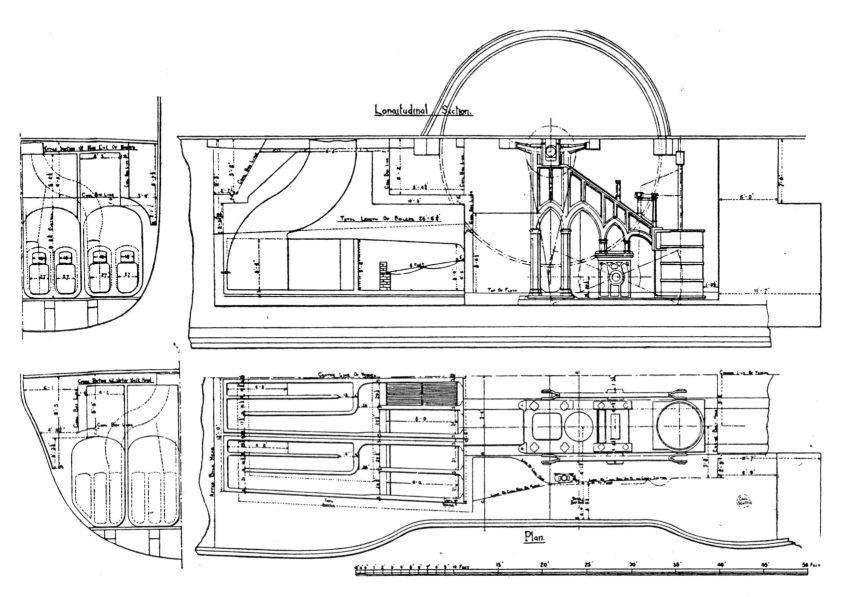

Machinery of *Hecla* and *Hecate*, 1830

A drawing showing the general arrangement of boilers, engine and paddle shaft.

(D K Brown collection)

Contemporary usage defined each cylinder as an 'engine' and hence multi-cylinder machinery took the plural – engines. For example, there is a letter of 9 April 1827 saying that *Dee* was to be fitted with 'two engines of 100HP each.'[24] (Today it would be described as a single 2-cylinder engine of 200HP.

Until about 1843 steam was generated in box boilers, rectangular tanks full of sea water through which flues carried the hot gases from coal burnt in three to four furnaces per boiler, boiling the water. The box was almost always of iron and lasted about three years before it had to be replaced at a cost of about £1500. A few copper boilers were used, though they were more expensive, as it was expected that they would last much longer. Performance in service, however, was disappointing and with the rising cost of copper and increasing steam pressure, iron boilers became universal.

Driver's boilers of 1842 were fairly typical. There were three boilers which had a combined weight of 40 tons and they contained 50 tons of seawater. Details of the boilers are given in the following table.

Table 35: PARTICULARS OF *DRIVER*'S BOILERS

Length x breadth x height	26ft x 9ft x 12ft 6in
Furnace (three per boiler)	7ft 8in x 2ft 6in
Flues	18in, 5in apart
Furnace to top of funnel	70ft

Twice during a watch, the bottom layer of water, which had the highest salt content, was blown out into the sea – a hazardous operation until the invention of the Kingston valve in 1837. Steam was used at 3-5lb/sq in to push the piston down the cylinder. It was not allowed to

A new design of direct acting engine, used in *Gorgon*

Designed by Seaward and Capel, this engine was powerful but caused severe vibration.
(Science Museum, London)

expand significantly, thereby wasting much of the energy in the steam. Machinery of this sort was heavy and weights for *Medea* are given below.

Table 36: MACHINERY WEIGHTS (*MEDEA*)

Engines	165 tons
Boilers	35 tons
Water in boilers	45 tons
Coal	320 tons
TOTAL	565 tons

The engines were also inefficient with coal consumption of up to one ton an hour or 110-190 tons for a ten-day passage.

During this first phase, up until 1837, the dominant engine builders were Maudslay and Boulton & Watt, who each built about a third of the engines while the other third was split between a further five builders.

Seward & Capel's engine of 1837 for *Gorgon* introduced a period of change in engine design. Only a few engines were built to this particular design but rival firms were inspired to develop engines which were powerful and relatively light and compact. All of them were direct-drive

Maudslay's 'Siamese' engines for
Devastation were very successful
(Science Museum, London)

engines, whereby the piston was joined to the crankshaft by a simple connecting rod, without a beam or side levers. Towards the end of the paddle warship era, Penn introduced the oscillating cylinder engine in which the cylinders were mounted on trunnions, through which the steam was admitted, and could move so that the crank was worked by an extension of the piston rod. At much the same date, Maudslay developed the Siamese engine which had two pairs of cylinders, each with a common connecting rod, arranged vertically below the crankshaft. The 1860 Committee on Marine Engines found that these two designs were the most satisfactory for paddle ships.[25]

Tubular boilers were introduced in 1843 in *Penelope* and *Firebrand*, the latter having 2250 tubes, each 2in diameter, ⅛in thick and 5ft 1in long. These carried the hot gases through the water, thereby increasing the heating surface. Such boilers had a cylindrical casing, worked at about 8lb/sq in and were considerably more costly than flue boilers. Despite this, their greater efficiency and the speed with which steam could be raised led to most ships being retro-fitted with tubular boilers during the 1840s. Brass tubes were generally employed, despite the risk of electrolytic action, as corrosion in iron tubes was too rapid.

Warship machinery design had diverged from that of merchant ships because of the need to keep engines below the waterline as far as possible for protection. This led to warship engines being designed with horizontal cylinders rather than the vertical ones more common in merchant ships.

When invited to tender, the engine builder would be given the weight, and the position of the centre of gravity of the machinery, together with a section of the engine room and a note of its length. The then Surveyor, Baldwin Walker, told the Committee of 1860 that, 'The general system adopted by the Admiralty for obtaining engines is, to specify beforehand the minimum area of fire-grate, and of absorbing surface of the boiler, on which the power mainly depends, and to approve of the size of cylinder, length of stroke, and other particulars of the engines before the tender is accepted'.[26] Guidelines (1860) recommend that the grate surface should be 0.68sq ft/NHP and that the heating surface should be 18sq ft/NHP. The engine builder would then be invited to fit in the most powerful engines available. Though NHP gave little indication of the actual power which could be delivered, it was a measure of engine size and therefore provided a reasonable guide to cost. As such it was quite often specified in a contract.

It is difficult from available accounts to decide whether the early engines were free from breakdowns or whether they were easily repaired; many accounts suggest that they were quite reliable, though Alexander Gordon (of Napiers), in a rather biased pamphlet of 1843 written to the Admiralty in an effort to acquire more work for his company, implies the opposite.[27] Boilers needed frequent attention and replacement and it is said that the Mediterranean mail packets needed boiler repairs every fourth voyage.

The Admiralty set up the 'Steam Factory' at Woolwich in 1836 to refit the machinery of the growing fleet of steamships. The Chief Engineer, Peter Ewart, was given the same salary as a Master Shipwright and under him and his successors the Factory made a major contribution to the steam fleet. It not only refitted machinery but designed, built and tested a wide range of new equipment, such as the Kingston valve, and also trained engineers for the fleet. Many of the engineers who founded Thameside engine works learnt their trade at Woolwich.

Only a few records dealing with the use of machinery have survived and these may not be typical. One such refers to *Terrible*, which in 1846 covered 15,731 miles at a mean speed of 6.83 knots of which only 6,277 were under sail alone.[28] In the Mediterranean, between March and October 1846, she covered 5,725 miles at 9.07 knots of which only 351 miles were under sail alone. However, in 1848, again in the Mediterranean, she sailed 64 per cent of the distance covered. Under steam, she travelled about five miles per ton of coal. It needs to be remembered that in the more far-flung parts of the Empire coal depots were few and far between and so it became essential to use sail.

At the very beginning of the steam era, engineers were part of the package supplied by the engine builder, though the Royal Navy soon built up its own engineering service, and engineers were given the rank of warrant officer, the traditional and respected rank of specialists such as the gunner. A number of seamen officers took courses to familiarise themselves with steam engines as a part of their general education, rather than to become engineers, and several books were published to the same end.

Paddle wheels

All the earliest paddles and the majority of those fitted to warships were the so-called 'common wheel'. With this type the boards, or floats, were fixed radially around the circumference. Typically, the wheel would be about 20ft in diameter and 6ft wide with some twenty floats, each 2ft 10in deep (radially) around the circumference. The paddles turned at about 18-20rpm. There was little that could go wrong with such a wheel and, if a float was damaged, it was easy to repair.

Table 37: PARTICULARS OF SOME TYPICAL PADDLE WHEELS

Ship	Date	Diameter		rpm	Notes and type
		ft	*ins*		*(common unless stated)*
Comet	1822	14	0	28	
Lightning	1823	17	9	25	Feathering; late in life
Meteor	1823	14	0	27	
Echo	1827	17	1	18	
Dee	1832	20	0	18	
Rhadamanthus	1832	20	0	18	
Acheron	1836	20	0	19	
Gorgon	1837	26	0	18	48 floats, 7ft wide
Cyclops	1839	26	0	16	
Scourge	1844	27	6	18	Floats 9ft x 2ft deep
Penelope	1843	32	0	16	
Terrible	1845	34	0	16	Floats 13ft wide
Furious	1850	27	4	16	

1861-2

It was soon realised that the performance of the wheel depended on its arrangement relative to the ship and to the water surface. Writing in 1860, Douglas gave some guidelines in which he suggested that the immersion of the float should be equal to its depth, that is the top of the lowest float should be just awash. It was also suggested that the diameter of the wheel should be 4½ times the stroke of the piston so that the inner edge of the board would have the same forward speed as the ship, and that the width should be a third of the diameter.[29]

Constant depth of immersion presented problems as the draught of

This feathering wheel was devised by Galloway in 1829.
Morgan, who bought the patent in 1830, much improved the design. The object was for the floats to enter the water smoothly, thus improving efficiency. *(Science Museum, London)*

a ship altered as coal was burnt, and the waterline further varied with speed as it altered according to the height of the bow wave. Wheels were later designed in which the floats could be drawn radially inwards but these 'reefing' wheels were complicated and it seems they were

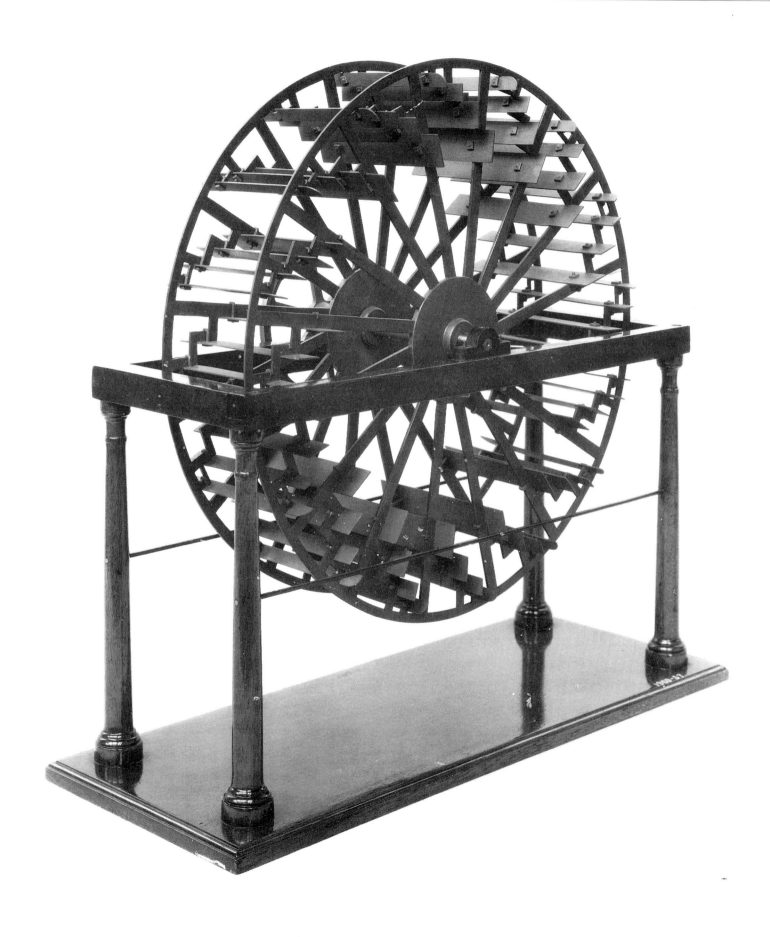

rarely used.

Reefing, however, did provide a solution to a different problem. Under sail, the drag of the immersed floats was considerable and it was usual to unbolt the lower floats for sailing. This was never an easy task and could be dangerous because, once removed, steam was not available if needed quickly. During *Rhadamanthus'* Atlantic crossing, stokers were rewarded for this unpleasant task with extra rum, and consumption was heavy.

That the float should move at the same speed as the ship is in accord with modern thinking. If the float slaps the water on entry, rather than sliding in gently, a considerable amount of power is wasted. The radial floats of a common wheel did so hit the water and in 1829 Elijah Galloway designed a feathering wheel in which a linkage to a crank on the wheel altered the angle of the floats so that they entered the water smoothly.

William Morgan, having bought the patent in 1830, improved the design, and in 1831 the Admiralty fitted a pair of Morgan wheels to the *Confiance* for comparative trials with her sister *Echo*, which had common wheels, and the feathering wheel proved considerably more efficient. Feathering wheels became universal in sea-going merchant ships but were rarely used in warships as it was feared that the linkage would be vulnerable to gunfire. Increasing the diameter of a common wheel reduces the shock loss on entry; a solution used on the Mississippi but impractical on sea-going ships.

The cycloidal wheel was another way of reducing power loss. Each float was split into several narrow strips, set stepwise, in advance of each other along a cycloidal curve, towards the rim. While it did not have all the benefits of the feathering wheel it was less prone to damage and was certainly better than the common wheel.

10. Paddle Steamer Trials

THE ADMIRALTY ORGANISED two major trials between pairs of generally similar ships in order to compare the performance of the screw propeller with that of the paddle. In fact, other advantages of screw propulsion were so great that it would probably have been chosen even if it had been less efficient.

For ships of this size, one would expect a modern propeller, in isolation, to have an efficiency of about 0.7 but this would be reduced behind a ship like *Rattler*, because of the interference between the propeller and hull flow, to something of the order of 0.53; a good modern feathering wheel might achieve an overall 0.5.

The trials between *Rattler* and *Alecto* need to be interpreted with caution as *Rattler*'s engines developed 360ihp as against 280 for *Alecto*.[30] The power put into the drive shaft would have been about 25 per cent less than these figures and may well have been different in each ship. This problem was recognised at the time and the Chief Engineer, Thomas Lloyd, had a thrust meter fitted to *Rattler* so that we at least have an accurate figure for her power delivered, but it was not possible to devise an equivalent scheme for paddle drive.[31]

If the trial results are corrected to equal power, *Rattler* had only a small advantage in efficiency as suggested by the figures above.[32] Under sail, *Rattler* merely had to turn her two-bladed propeller so that it lay up and down behind the stern post whilst *Alecto* had to remove her paddle boards; and even after all this, *Rattler* still sailed faster. The screw is better able to cope with varying loads and speeds and hence *Rattler* had the advantage in the towing trials.

In the 1849 trials between *Niger* (screw) and *Basilisk* (paddle) the latter developed 1032ihp and *Niger* only 790ihp. At equal power, there was no measurable difference in speed. However, the trial report brought out the real advantages of the screw. *Niger*'s hull was 95 tons lighter than that of *Basilisk* as the screw and its shaft needed little extra support, while massive supports were required for the wheels of *Basilisk*. The screw machinery was much better protected as it was well below the waterline while the top of the paddle machinery was some 6ft above the water.

When a paddler rolls in a seaway, first one and then the other wheel is more deeply immersed and the uneven thrust causes the ship to yaw. At the time, it was feared that damage to one wheel would immobilise a ship but the two recorded examples of this type of damage suggest that such fears were much exaggerated. The French *Fulton* had one wheel smashed in the Parana River but managed to go down the river and then return upstream, still with only one wheel, towing some barges. *Valorous* had a wheel badly damaged off Sevastopol but apart from severe vibration she was little affected. The efficiency of a paddle wheel is not determined by the width of the wheel and so even the loss of half the combined width of two wheels has little effect. Finally, the screw had the advantage of not being seen; with the funnel lowered, a screw vessel could look like a sailing ship, and win over some of the diehards, offended by the appearance of paddle vessels.

Speed trials of each new design of paddle steamer were carried out with care and it was recognised by most people that the average of runs in opposite directions should be used to cancel the effect of tide and current. However, there were a number of factors which could not be fully controlled but which affected the accuracy of the trial results. Captain Henderson of the *Gorgon* pointed out that cleaning the copper sheathing could increase speed by 1½ knots, whilst housing the topmasts increased speed by up to 3 knots in a head wind. The 1860 committee were told of others:

'… coal of "average" quality is a vague definition, but this is not all. It is well known that the actual power developed by an engine, which is seldom two days alike, depends on many circumstances other than its own merits and those of the coal and boiler; such as the draught of water of the ship, state of the weather, direction of the wind as affecting the ventilation of the stoke-hole and the draught of the boiler, skill and energy of the stokers, duration of trial, quantity of water blown out of the boilers to prevent encrustation, degree of expansion used etc…'

Quoted trial speeds represent about the best speeds which could be achieved in favourable circumstances, but, for the reasons given above, they are unlikely to be accurate to better than one quarter knot. Figures given to three decimal places are a quirk of the averaging process and mean little. Indicators were quite accurate and hence so are the ihp figures given, but there was no measurement of power applied to the paddle wheel, about 70-75 per cent of the ihp.

Phoenix shown in her later days with three masts. The notes on the drawing relate to a proposal of 1853 to reduce the height of the lower masts, presumably to improve her speed under steam.

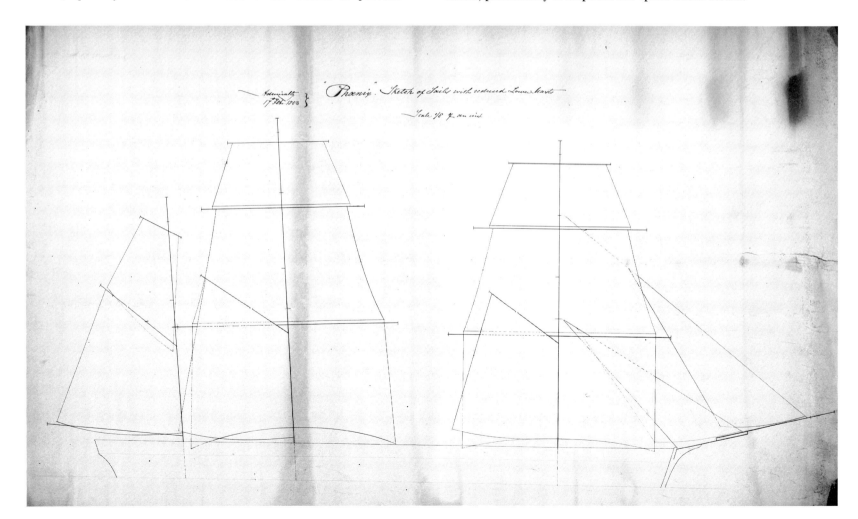

11. Sailing Characteristics and Seaworthiness

PERFORMANCE UNDER SAIL was not the *raison d'etre* of these vessels but the remarks quoted in Part I suggest that their ability to sail did not upset their Captains. The Surveyor recognised in evidence to the 1860 Committee that finer lines were desirable but explained that fine ends meant insufficient buoyancy to support their weight. This meant that the buoyancy had to be added nearer amidships, adding to bending stress and tempting the Board to add weapons, negating the purpose of the change and adding to cost – a problem not unknown today.

It was difficult to arrange a mast passing through the machinery spaces and, in consequence, most early steamships had two masts, one either end of the machinery. This rig was not well balanced for sailing and, in the late 1830s, many had a mizzen mast fitted, the main mast being moved a little forward if possible. It was also suggested that the three-masted rig, with spars which were smaller individually, had a smaller air resistance when under steam.

Tables of mast and spar dimensions and of sail area follow.

Table 38: MAST AND SPAR DIMENSIONS OF PADDLE WARSHIPS (as standardised in Nov 1847)

Category MASTS		1 Length ft in	Dia in	2 Length ft in	Dia in	3 Length ft in	Dia in	4 Length ft in	Dia in	5 Length ft in	Dia in	6 Length ft in	Dia in	7 Length ft in	Dia in	8 Length ft in	Dia in
Main	To hounds	66 0	30½	61 0	28	57 0	25½	53 0	23½	48 6	21½	45 0	19½	41 6	18	38 6	17
	Head	16 6		15 6		14 6		13 6		13 6		11 6		10 6		10 6	
	topmast, hounds	55 6	15½	49 6	14	44 0	12½	39 6	11½	36 0	10½	32 6	9½	30 0	8½	27 0	8
	head	7 3		6 6		6 0		5 6		5 0		4 6		4 0		3 9	
	t'gallant, stops	21 3	8	20 0	7½	19 0	7	18	6½	17 0	6½	16 0	6	15 0	5½	14 0	5½
	pole	15 0		14 0		13 0		12 3		11 6		10 9		10 0		9 6	
Fore	To hounds	61 0	30½	57 0	28	53 0	25½	48 6	25½	45 0	21½	41 6	19½	38 6	18	35 0	17
	head	15 6		14 6		13 6		12 6		11 6		10 6		10 0		9 6	
	topmast to hounds	54 0	9½	48 0	9	42 6	8½	37 6	8	33 0	11½	29 6	7	26 0	6½	23 0	6
	head	10 0		9 0		8 0		7 0		6 3		5 6		5 0		4 6	
	t'gallant, stops pole	As Main								As Main							
Mizzen	To hounds	54 6	22	51 0	20	47 6	18	44 6	18½	41 6	15	38 6	13½	36 0	12½	33 6	11½
	head	9 6		9 0		8 6		8 0		7 6		7 0		6 6		6 3	
Bowsprit	ex housing	30 6	26½	28 6	24	26 6	22	24 9	20	23 0	18	21 6	16½	20 0	15	18 9	13½
	Jib-boom	47 3	16	42 6	14	38 0	12	34 0	10½	30 6	9	27 6	8	24 6	7	22 0	6

Table 39: MAST AND SPAR DIMENSIONS OF PADDLE WARSHIPS (as standardised in Nov 1847)

Category YARDS		1 Length ft in	Dia in	2 Length ft in	Dia in	3 Length ft in	Dia in	4 Length ft in	Dia in	5 Length ft in	Dia in	6 Length ft in	Dia in	7 Length ft in	Dia in	8 Dia in
Main	lower yard	87 0	27½	82	18½	74 0	16½	67 0	15	61 0	13½	56 0	12½	52 0	11½	48 0	10½
& fore	tops'l yard	59 0	13	55 0	12	51 6	11	48 0	10	45 0	9	42 0	8½	39 0	8	36 6	7½
	to'gallant yard	41 0	8½	38 6	8	36 0	7½	33 6	7	31 0	6½	29 0	6	26 6	5½	25 0	5
	royal	30 0	5½	28 0	5	26 6	4½	25 0	4	26 6	3½	21 9	3¼	20 3	3	19 0	2¾
	gaff	49 6	13	45 0	12	41 6	11	37 6	10	34 0	9	31 0	8½	28 0	8	25 6	7½
Mizzen	gaff tops'l	17 3	4¾	16 0	4½	15 0	4¼	14 0	4	13 0	4	12 0	3¾	11 0	3½	10 6	3½
	boom	52 0	12	47 0	11	42 6	10	38 6	9½	34 6	9	31 0	8½	28 0	8	25 3	7½
	gaff	35 0	8	32 0	7½	29 0	7	26 6	6½	24 3	6	22 0	5½	20 0	5	18 6	4½
Trysail	gaff (Main & fore)	37 0	11	33 6	10	30 0	9	27 0	8	24 6	7½	22 0	7	20 0	6½	18 0	6

Table 40: SAIL AREAS OF PADDLE WARSHIPS (1847) (square feet)

Category		1	2	3	4	5	6	7	8
Main	courses	4342	3730	3080	2550	2100	1785	1490	1300
	topsails	2819	2325	1938	1577	1300	1108	934	765
	to'gallants	946	825	727	640	562	490	420	362
	royals	489	425	370	320	278	242	209	182
	gaff sails	3504	2900	2400	1980	1640	1360	1125	931
Fore	courses	3915	3350	2780	2300	1910	1600	1370	1160
	topsails	2819	2325	1938	1577	1300	1108	934	765
	to'gallant	946	825	727	640	562	490	420	562
	royals	489	425	310	320	278	242	209	182
	gaff sails	3357	2780	3210	1920	1590	1325	1100	910
Mizzen	gaff sail	2104	1750	1470	1225	1025	860	717	598
	gaff tops'l	848	680	943	435	350	280	223	179
Outer jib		1880	1530	1248	1001	805	650	532	430
Inner jib		1033	863	714	589	496	412	334	280
Fore staysail		1181	992	821	663	554	459	386	324
TOTAL		30672	25725	21435	17737	14750	12411	10403	8730

Table 41: STABILITY OF *VALOROUS*

Condition	Displacement (tons)	Draught, mean ft in	Metacentric height (ft)
Arctic deep	2373	18 2	2.47
Deep	2243	16 5	2.16
Light	1851	14 2	1.16

openings – such as hatches and doors – strong enough to withstand the worst seas. The very active usage of these warships together with the small number lost suggests that they were indeed 'seaworthy'.

Seakeeping and seaworthiness

A number of subjective comments on seakeeping have been quoted in Part I but their value is limited. Modern studies on the writing of questionnaires show that the pride of a captain in his ship will ensure a favourable reply unless it is truly awful. It is interesting that *Hermes* always attracted unfavourable remarks whilst near-identical ships were praised, this possibly being an example of giving a dog a bad name. Seakeeping is a complex 'umbrella' heading involving almost unrelated attributes such as pitch, heave, roll, wetness, slamming etc, some of which are hard to estimate with human sense organs.

The ships discussed were all of fairly deep draught and, at the speeds they could achieve, slamming should not have been a problem. Similarly, freeboard was adequate to prevent undue wetness.[33] Pitch and heave for ships of generally similar form would depend mainly on length and there is no reason to suppose that these ships differed greatly in behaviour from sailing ships of the era. Most, if not all, lacked bilge keels and their light rig would have provided less roll damping than that of the traditional sailing ship. It is interesting that the word 'easy' occurs frequently in accounts of their behaviour in a seaway. Words and their usage change meaning but, today, easy would imply a slow, long-period roll suggesting a fairly small metacentric height but figures have not been found to support this hypothesis.

The only good data on the stability of paddle warships is for *Valorous* and comes from W E Smith's work book for 1875, when she was very old.[34]

These data are insufficient to match against modern safety criteria but, with a good freeboard, they would probably seem adequate today.

Seaworthiness adds to seakeeping the concept of a hull and its

These sketches of *Phoenix* show how the rig was altered depending on wind direction.

In some directions sail could assist the engines but, head to wind, the topmasts were lowered to reduce aerodynamic drag.

HER MAJESTY'S STEAM SHIP, PHŒNIX.

Plate 28.

4 points from the Wind

5½ points from the Wind

Head to the Wind

Before the Wind

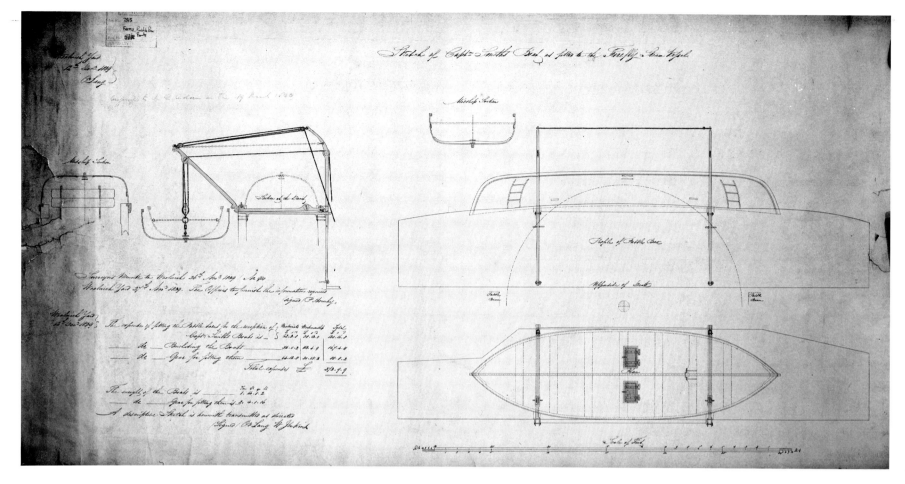

12. Armament and fittings

THE GREAT MAJORITY of paddle steamers carried all their guns on the upper deck and it was very easy to alter the outfit. In consequence, changes were frequent and it should not be assumed that at any one time all ships of a class had the same armament. Viewed overall, despite these differences, there was a general uniformity of approach and, to some extent, the actual outfit must have reflected the availability of guns in the armament yards.

The early, small steamers of the 1820s were built as tugs and tenders and, as such, they mounted two or three 6-pounders. Around 1830, as their military potential was realised, they were given four or even six 18-pounders. The bigger ships of the late 1820s and early 1830s had an armament which remained fairly standard for second and third class sloops, with a long 32-pounder pivot forward and a shorter 32-pounder on either beam (two in bigger ships). The pivot was either a 42cwt or 25cwt gun and the broadside weapons were either the 25cwt (sometimes called a 'gunnade') or the 17cwt carronade.

By the late 1830s, attempts were made to offset the small number of guns which could be carried on a paddler by mounting very large

Paddle Box Boats. This diagram shows the ingenious way in which paddle box boats were stowed and launched.

These big boats, carrying up to 100 men, were an important part of the armament in the coastal operations where many paddlers spent much of their service. The boat was stowed upside down on a cradle above extended paddle boxes. To hoist out, the cradle was lifted by inboard davits and then itself formed the davits to lower the boat into the sea.

guns. The *Medea*s mounted one or two 10in/84cwt guns as pivots or, in some cases, the 8in/52cwt in place of one of the 10in. This use of very large pivot guns persisted and by the 1840s some ships received the more effective 8in/65cwt or the more accurate 68pdr/95cwt, perhaps the best gun available. Later, in the 1860s, some ships had the 110pdr/82cwt RBL or the smaller but more useful 40pdr/35cwt RBL.

The enthusiasts for these big guns claimed that a paddler, so armed, could stand off from a sailing battleship and, at least in a calm, destroy it at long range. They forgot that the battleship also carried big pivot guns and was, in addition, a steadier gun platform. Furthermore, the battleship could use her boats to turn her, opening up the full broadside. In any seaway the effective range of all guns was quite short as training and elevating were very slow. In the light of these claims for long-range firing it is surprising that so many pivot guns were the inaccurate, shell firing, 8in and 10in models, inaccurate because the fuse was an unbalanced weight which meant that the spherical shell did not fly true.

The armament of the *Gorgon/Cyclops* group followed a similar pattern, though the size of the broadside guns was increased considerably. In the 1860s a typical first class sloop would mount a 68-pounder and a 10in pivot and four 32pdr/42cwt on the broadside. The second class frigates varied even more but a common outfit was two 10in pivots and four 68-pounders on the broadside.

The first class frigates varied so much in size and design that one can only refer the reader to Part I and Table 30. Even though they carried more guns, the tendency to mount the largest possible size was still noteworthy.

Since the policy of few big guns was forced on the paddlers by their obstructed broadside, it is amusing to note that the early screw ships, with a full broadside, were criticised because they could not carry a pivot aft since that site was occupied by the propeller lifting trunk and the fine lines of the stern did not offer much support.

Boats

Many paddle steamships were employed in suppressing the slave trade and piracy, as colonial policemen and in assisting friendly governments. They were also often used in survey work and in all these tasks the ships' boats played an essential part. The landing of armed parties was an important operation and the boats may well be seen as a part of the ship's armament.

Barracouta's outfit (1847) may be taken as typical and she carried one each of the following:

> 25ft cutter
> 27ft whaler
> 18ft gig
> 22ft cutter
> 14ft dinghy
> 30ft steam pinnace, carried over the engine room.

She also carried two box boats, invented by Captain Smith in about 1830. These boats carried about one hundred men for landing parties and were stowed, upside down, on the paddle boxes which the boats were shaped to fit. (The illustration and caption on page 78 explain how these boats were worked)

13. Builders

ALL BUT TWO of the wooden hulled vessels were built in the Royal Dockyards whilst the iron ships were built by commercial yards. The great majority of the early steam ships were built at Woolwich, partly because Lang was Master Shipwright there and partly because Woolwich was convenient for the leading engine builders, most of whom worked on the Thames.

Pembroke was the leading builder of the later ships since that yard was created as a specialist building facility.

Table 42 lists costs of hull and machinery for some vessels. The figure for machinery, which was built by contract, was the price paid. The cost given for the hull of dockyard-built ships is calculated on a rather different basis. It would include the cost of the raw material, timber, iron etc and the cost of the wages of those directly employed on building the ship. The cost of providing the building slip, workshops and other dockyard facilities is not included and nor is the cost of indirect staff – managers, police etc. There was nothing dishonest in this breakdown; the 'missing' items were declared in other Dockyard costs and Parliament was well aware of what was shown, but it does make comparison with commercial builders difficult.

Table 42: SOME TYPICAL COSTS

Ship	Date	Cost in £		
		Hull	Machinery	Total
Comet	1822	4314	5050	9,364
Lightning	1823			14,661
Dee	1832			27,700
Gorgon	1837			54,306
Devastation	1841			42,168
Sphynx	1846	25,000	25,000	50,000
Gladiator	1844			53,165
Furious	1850			59,323
Terrible	1845			94,650

STORIES PERSIST TO this day that the Admiralty and senior officers were opposed to steamships yet, at a time when most of the sailing fleet was in reserve, the steamships were all in active commission, except for those actually refitting which usually meant having new boilers installed. As has been said before, there were frequent demands from Admirals at sea for more steamships.

It is less clear how often they used their engines; long passages had to be under sail for much of the time as they could not carry enough coal to feed their inefficient engines for long periods. At home coal was cheap, about five shillings per ton, but in remote areas the cost was about four times as much and depots were few and far between. No captain would wish to find himself without fuel and some degree of economy was quite reasonable. There certainly was a pride in handling these ships under sail and some captains would see the use of engines as a last resort. The few records of engine usage are probably not typical but do suggest that considerable use was made of steam.

Opinions on steamships tended to extremes, there were certainly some, not very influential, who were opposed to the steamship but a greater number over-valued the steamship armament of a few large guns, claiming that a paddle sloop was the equal of a sailing battleship. The earliest steamships were built as tugs to tow sailing battleships into action and this remained an important operational task. In the bombardment of Sevastopol in 1855, the sailing battleships each had a steamship lashed alongside to bring them into action and then take them out again.

Modern writers tend to undervalue the paddle warship, assuming that since it was replaced by the screw ship, the paddler must have been grossly inferior. This was not the case. Though the screw ship always had the edge, its advantage was quite small at first. There were a number of problems to be solved before the screw ship could establish a clear superiority. There was no way to design a propeller until Froude's work of about 1872, and earlier screws were selected on the basis of trial and error, which meant that several expensive propellers had to be made and tried, also at considerable expense, before a usable design could be arrived at.

The stern gland caused problems. Rapid wear and leakage, for example, almost caused the loss of the battleship *Royal Albert* in 1856, a problem eventually solved by Penn using lignum vitae. Vibration was even worse in the screw ship than in paddlers and careful trials in 1854 between *Himalaya* (screw) and *Victoria and Albert* (paddle) led to the choice of paddles for the new Royal Yacht *Osborne* as they caused less vibration.

Those who see the Admiralty as slow in adopting the screw should look at Samuel Cunard, who built his last transatlantic mail steamer *Scotia* with paddles in 1861, and she was only converted to screw in 1879.

Though paddle warships fought no major battles, they were involved in innumerable minor campaigns against slavers, pirates, rebels and other enemies of the Queen. In all of these they proved effective and reliable; above all, they taught the Royal Navy about the certain mobility of steam.

Appendix

Lightning, the first Post Office steam packet.

She was brought into the RN in 1837 under the name *Monkey*. Her numerous and lengthy renamings are given in the text. She is often confused with the RN's *Lightning*.

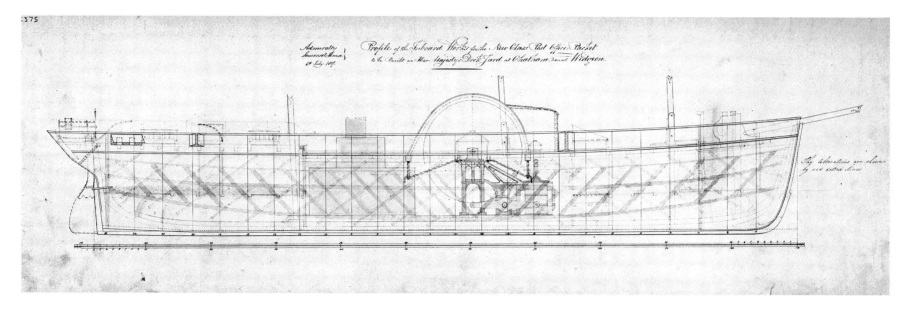

Post Office Packets

AT THE END of the Napoleonic Wars, mail was carried along the sea routes of the British Isles by privately-owned vessels which, by 1816, included steamships. In 1816 it was decided that the Post Office should operate its own fleet of ships and, it would appear, it was also decided that they should all be steamships. The first two were *Lightning*, described below, and the slightly smaller *Meteor*, which entered service in 1821. The following year an even older steamship, the *Ivanhoe*, built in 1820, was purchased, thereby becoming the oldest vessel in the fleet. Three sailing packets were taken over in 1821 and these were used for some time as colliers.

The Post Office used a number of designers, including Lang and Symonds, and many builders, most of whom were not involved in Admiralty work. Some, perhaps all, were built under Admiralty supervision. These packets differed from warships in having two passenger cabins, one for ladies, the other for men. Both had two-tier bunks. They were vehicle ferries, too, carrying coaches on the upper deck. Most had armament allocated though it is not clear if it was fitted. They had black hulls, usually with a broad white band.

Routes

The main routes operated, with approximate dates were:

> Weymouth – Channel Islands (initially Southampton): 1827-45
> Holyhead – Howth (Later Dublin): 1821-49
> Milford – Waterford (initially Dunmore): 1824-53
> Port Patrick – Donaghadie: 1825-45
> Dover – Calais (Ostend was added later): 1826-54

Widgeon, one of the fastest mail packets.

She was used to race against the *Archimedes*, comparing paddle and screw propulsion, and though *Widgeon* was slightly the faster, corrected for her extra power and smaller size, it was clear that the advantage lay with the screw.

Numbers

There were a considerable number of packets and the table below shows the numbers built each year and the cumulative total.

Table 43

Year	Built	Total	Year	Built	Total	Year	Built	Total
1821	5	5	1828	–	24	1837	3	34
1822	1	6	1829	1	25	1838	2	36
1823	2	8	1830	–	25	1840	2	38
1824	3	11	1831	4	29	1844	1	39
1825	3	14	1832	–	29	1845	2	41
1826	4	18	1833	1	30	1846	1	42
1827	6	24	1834	1	31	1847	4	46
						1848	2	48

In 1835, it was said that the packets had cost a total of £273,018 4s 9d to build.

The Admiralty takeover

By 1836 there was increasing dissatisfaction with the service provided by the Post Office and it was said that the ships were poorly maintained, sometimes badly constructed, and that no effort was made to keep them up to date. In consequence, it was decided that the service should be taken over by the Admiralty, who were already running the Mediterranean mail service, and all existing packets were brought into the Royal Navy in 1837.

Several of the Post Office ships had names already in use by warships

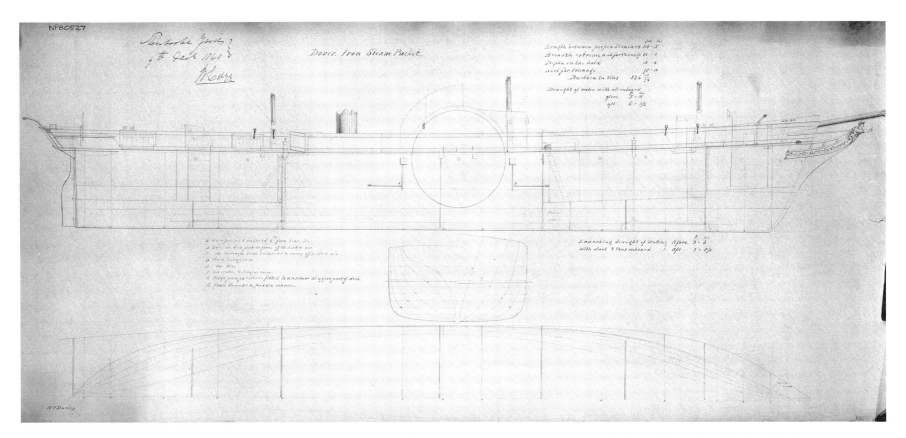

Dover, a mail packet and the first iron ship ordered by the Admiralty.

Iron ships could not go out of sight of land until Airy developed a means of correcting the compass for use in such a hull in 1838. The Admiralty kept careful records of her maintenance costs for comparison with wooden hulls but the results were inconclusive in the early years.

and so the whole fleet was renamed. Initially, the Admiralty welcomed these additional ships, seeing them as a wartime reserve but, before long, it was realised that they were a drain on scarce manpower and on maintenance effort and by 1859 most services had been handed over to contractors. Only a few of the ships were sold and most remained in naval service as tugs, tenders and yachts. One may wonder where the savings came from with the ships still requiring crews and maintenance and whether the new tasks were really necessary.

Table 43 lists all the packets, including those lost before being taken over.

Some interesting packets
The following vessels are chosen because of their special interest and hence should not be seen as typical.

Lightning, one of the Post Office's first two steamers, was designed and built by Elias Evans of Rotherhithe, well known for his early steamships, and built under Admiralty supervision. Her wooden hull was designed using Sepping's diagonal framing and cost £4,861. She had Boulton & Watt side lever engines of 80 NHP which cost £5,608. While in Admiralty service her engines were first altered, then replaced and she was reboilered. In 1851 she had 130 NHP engines of 373 ihp driving 14ft wheels at 27rpm.

There were six bunks in the ladies' cabin and fourteen for men. She had two masts with fore and aft rig, and a crew of about thirteen. Her funnel was tall and thin, even by the standards of the day, until a more

normal one replaced it after she was converted to a tug.

She left her builder's yard on 16 May 1821 and ran into a series of gales, only reaching Holyhead on the 29th. However, the actual passage only took five days and it was thought that the gales would have defeated a sailing ship of any size. By the end of 1821 she had made 140 crossings of the Irish Sea, far more than a sailing ship could have managed.

On 12 August 1821 she carried King George IV to Ireland after the Royal Yacht became becalmed. To mark the occasion she was renamed *Royal Sovereign, King George The Fourth*, which was shortened by the end of the year to *Royal Sovereign* and later to *Sovereign*; in 1837 the Navy gave her the rather less prestigious name of *Monkey*. In 1839 she was converted to a tug and was used as such until sold in 1888.

Widgeon is of interest as, in the spring of 1840, she was used in comparative trials with Pettit Smith's prototype screw steamer *Archimedes. Widgeon* was the fastest packet in the service and, though she was slightly faster than *Archimedes*, the trial was rightly seen as a victory for the screw since *Widgeon* had slightly more power and was of considerably less displacement.

Widgeon was designed by Symonds and built by Chatham Dockyard.

She was laid down in June 1837, launched in September, and entered service at the end of the year, which showed how fast Dockyards could work when it was necessary. She had a Seaward & Capel engine of 90 NHP (said to be of 190 ihp) driving common wheels at 26rpm for a speed in service of 10-10½ knots. The hull cost £4,251 and the machinery £5,065.

She was rated as a tug from 1848 but does not seem to have been much used. She was sold in 1884 for £460.

Dover was the first iron-hulled ship ordered by the Admiralty and was ordered from Laird in 1840. In appearance she was a typical: a small ship of the day with a clipper bow, figurehead, decorated transom and circular paddle boxes. She had three masts and was rigged as a topsail schooner; a funnel was positioned behind the paddles. There were the usual ladies' and mens' cabins aft and another small passenger space forward.

She had a Fawcett side lever engine of 90 NHP, tubular boilers and her 13ft 6in wheel gave a speed of 10½ knots at 28rpm.

She ran on the cross-Channel service until 1847, when she was refitted and given a small gun for service on the River Gambia where she remained, with a short break, until 1866 when she was presented to local merchants.

Last and fastest

In 1846 there was a design competition for four large and very fast paddler packets for the Holyhead run. The ships and their designers

Caradoc (Symonds) was much slower than *Banshee*, by about 20-40 minutes in a four-hour crossing. It is probable that the difference was in the power delivered by the engines rather than in form.

Both *Banshee* and *Caradoc* were commissioned as gunboats during the Crimean War.

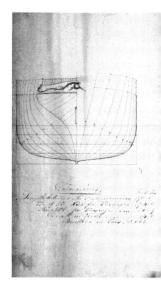

The fastest. In 1846 there was a competition for the design of fast packets for the Holyhead route.

Banshee, designed by Lang, was consistently the fastest in service, recording the fastest passage in both 1848 and 1849 and the best average speed over the two years. She reached 16.13 knots on trial.

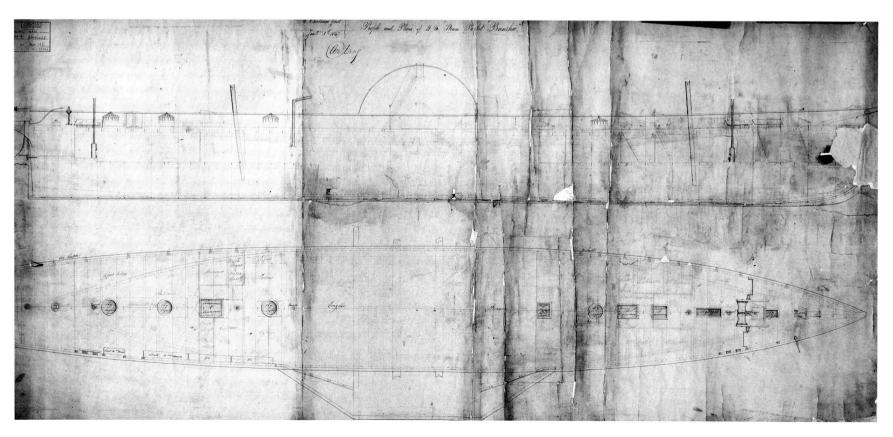

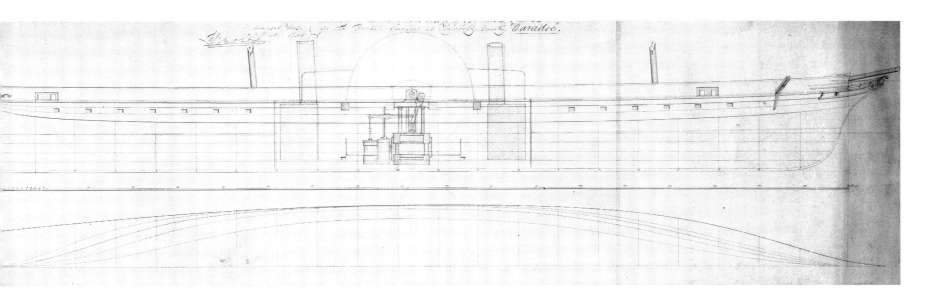

Table 44: POST OFFICE PACKETS

STATION RN Name	PO Name	Tons	Length ft in	Beam ft in	Depth ft in	Draught Fr'd ft in	Aft ft in	Builder	Launch	Notes
DOVER										
Ariel	*Arrow*	149	107 11	17 3	10 3	–		–	–	76NHP
Charon	*Crusader*	125	100 0	16 2	9 11	6 2	6 10	–	1827	
Swallow	*Ferret*	133	107 6	15 11	9 9	9 9	8 9	Pitcher	1831	70NHP
Beaver	*Salamander*	128	102 2	16 2	9 11	6 7	5 10	–	1827	62NHP
Myrtle	*Firefly*	116	96 3	16 1	9 7½	6 7	9 10	Fletcher	1831	50NHP
Widgeon	*Widgeon*	164	180 0	10 10	–	6 6	6 6	Chatham	1837	90NHP
Dover	*Dover*	224	110 5	21 0	10 6	5 2½	5 2½	Laird	1840	197NHP
	Garland	295	140 0	21 0	–	–		Fletcher	1846	
	Onyx	292	139 0	21 0	–			Mare		
	Princess Alice	270	140 0	20 6	11 0			Mare		Iron
	Scout									
	Undine	284								Iron
	Violet	292								Iron
	Vivid	352	150 0	22 0						
WEYMOUTH										
Fearless	*Flamer*	165	112 6	18 5	11 1			Fox	1876	
Wildfire	*Watersprite*	186	116 6	18 5	11 7			Graham	1826	
Dasher	*Dasher*	260	120 1	21 8	13 0			Chatham	1837	
	Cuckoo									
	Ivanhoe									
	Meteor	296	126 0	33 0						
	Pluto									

Table 44: Post Office Packets *(continued)*

STATION RN Name	PO Name	Tons	Length ft in	Beam ft in	Depth ft in	Draught Fr'd ft in	Aft ft in	Builder	Launch	Notes
MILFORD HAVEN										
Jasper	*Aladdin*	230	114 7	38 4	12 5			Simmonds	1824	100NHP
Adder	*Crocodile*	236	116 0	20 10	12 0			Graham	1826	85NHP
Pigmy	*Sibyl*	227	114 6	20 9	13 2			Humble	1827	80NHP
Advice	*Vixen*	175	108 4	20 0	11 6			Deptford	1827	
Prospero	*Belfast*	288	129 6	19 10	11 5			Belfast	–	161NHP
LIVERPOOL										
Kite	*Aetna*	300	124 11	22 8	14 5			} Humble & Hurry	{ 1825	152NHP
Lucifer	*Comet*	387	155 3	22 7½	14 1				1825	190NHP
Shearwater	*Dolphin*	343	136 11	22 8½	14 9			Graham	1826	160NHP
Redwing	*Richmond*	139	144 4½	15 9	10 8			Hunter	1834	60NHP
Avon	*Thetis*	361	144 0	22 9	14 7			Graham	1825	160NHP
Monkey	*R Sovereign*	211	106 3	21 0	11 4			Evans	1821	101NHP
Urgent	*Collonsay*	561	170 7	26 0	17 8			Purchased	1837	291NHP
Merlin *Medusa* *Medina* }		889	125 0	33 0	16 5			Pembroke	1838 1838 1840	300NHP
HOLYHEAD										
Cuckoo	*Cinderella*	234	120 6	20 0	12 6			Green	–	100NHP
Zephyr	*Dragon*	237	106 0	20 10	12 3			Graham		104NHP
Doterel	*Escape*	237	118 7	20 9	12 7			Graham		100NHP
Gleaner	*Gulnare*	351	138 0	23 0	13 0			Chatham	1833	130NHP
Sprightly	*Harlequin*	234	119 9	21 1	12 8			Green		100NHP
Otter	*Wizard*	237	120 0	20 9	12 5			Graham		100NHP
PORT PATRICK										
Asp	*Fury*	112	89 7	16 3	8 11					50NHP
Pike	*Spitfire*	112	84 0	16 3	8 11					50NHP

Table 45: 1846 HOLYHEAD FERRIES
Summary of Particulars

Name	Designer	Launch	Tons	Length ft in	Beam ft in	Depth ft in	NHP	Fate
Caradoc	Symonds	1 Mar 1847	662	193 0	26 9	14 9	350	Sold 12 May 1870
Banshee	Lang	1847	670	189 0	27 2	14 9	350	BU 1864
Llewellen	Miller Ravenhill	1847	654	190 0	26 6	–	350	Sold 1850
St Columba	Laird	1847	719	198 6½	27 3	15 5	350	Sold 1850

were: *Banshee* (Lang), *Caradoc* (Symonds), *Llewellen* (Miller and Ravenhill) and *St Columba* (Laird). Particulars are given in Table 44. It is surprising that, since they were intended to test different design styles, no formal trials were carried out and even basic data such as ihp were not recorded. (*Caradoc* is said to have had 997 ihp)

They all had engines of 350 NHP but it is likely that their real power differed considerably. In service, *Banshee* was clearly the fastest, making 16.13 knots on trial, recording the fastest passage in both 1848 and 1849, as well as the shortest average time over the two years. *Llewellen* was never far behind but the other two were much slower, taking 20-40 minutes longer in a four-hour crossing. The service was sold in 1848 to the City of Dublin Steam Packet Company and *Llewellen* and *St Columba* were sold to that company. *Banshee* and *Caradoc* were given three 18-pounders and commissioned for Mediterranean service, where they were active during the Crimean War as gunboats and despatch vessels. *Banshee* was broken up in 1865 and *Caradoc* in 1870.

Notes

[1] E C Cuff. 'Naval Inventions of Charles, third Earl Stanhope, 1753-1816', *The Mariners Mirror*, vol 66, 1942.

[2] D K Brown. *Before the Ironclad*, London (1990).

[3] Nominal Horse Power (NHP) was a measure of the geometry of the engine and bore little relation to the real power developed. It was defined as:

$$NHP = \frac{7 \times \text{Area of piston} \times \text{equivalent piston speed}}{33,000}$$

where equivalent piston speed is a function of the stroke equal to 129.7 x (stroke)⅓.35.

[4] See Part II, section 18, for trials accuracy.

[5] PRO (Public Records Office) ADM 95/87

[6] H N Sulivan. *Life and Letters of Admiral Sir B J Sulivan*, London (1896)

[7] P W Brock and B Greenhill, *Steam and Sail*, Newton Abbot (1973)

[8] PRO ADM 95/87

[9] PRO ADM 95/87

[10] PRO ADM 95/87

[11] PRO 95/880

[12] PRO ADM 95/88

[13] PRO ADM 95/87

[14] PRO ADM 95/87

[15] See Part II, Chapter 13 and Table 42 for costs.

[16] D Griffiths. *Brunel's Great Western*, Wellingborough (1985)

[17] Select committee on Estimates, 1847-48, Parliamentary Papers, Vol XXXI, Part I. (Future references to '1847 Committee') Evidence by W H Henderson.

[18] J A Sharpe, *Memoirs of Rear Admiral Sir William Symonds*, London (1858)

[19] As 2.

[20] J Fincham. *A History of Naval Architecture* (1851). Reprinted London (1979)

[21] H Challoner. 'The Historical River Steamer *Pioneer*'. *Auckland Historical Journal*, vol 2 no 1, October 1963.

Part II

[22] J F Coates, 'Hogging and breaking of frame built wooden ships', *The Mariners Mirror*, Vol 71 (1985)

[23] As 2.

[24] PRO ADM 129 Apr 1827

[25] Committee on Marine Engines 1860, (held in Naval Library)

[26] Ibid.

[27] A Gordon, *Marine Steam Engines of the Royal Navy*, Glasgow (1843)

[28] E P Halsted. *Steam Fleet of the Navy*. London, (1850).

[29] H Douglas, *Naval Gunnery*, London (1855)

[30] As 2.

[31] In 1866 William Froude and Henry Brunel devised a dynamometer for the paddle engines of *Great Eastern*. This seems to be the only case of measurement of output from paddle engines and results have not been found.

[32] D K Brown. 'The Introduction of the Screw Propeller into the Royal Navy'. *The Naval Architect*, March 1976.

[33] A R J M Lloyd, *Seakeeping – ship behaviour in rough weather*, Chichester (1989)

[34] W E Smith, manuscript work book held in the National Maritime Museum.

[35] PRO ADM 593/5

Sources

Admiralty Collection, National Maritime Museum
(Held in the Brass Foundry, Woolwich Arsenal)

EARLY SHIPS
Lightning   1822 L/X; A/B; P/G; 1833 LP/G/W
Meteor   1837; U

Alban class
Alban   1823; L/P/G/U 1835; S 1840 (Lengthened) L/P/G/U/X 1847; S
African
Echo   1833; X 1836; G/U as tug G
Carron
Confiance   1836; G/U
Columbia   1834 as troopship LP/G/U
Dee   1827; L 1838; L/G/U 1841-2 as troopship P/O/G/U
Firebrand   1831; L/P/G/U
Flamer   L/C 1837; P/G/U 1850; P/G/U
Black Eagle   1844; L
Pluto   1831; L/P/G/U/S 1835; G 1851; S 1856; P/G/U
Firefly   1832; LP 1836; S
Spitfire   Outline; P D X
Blazer   A/B P/G/U 1847; S
Tartarus   1833; L/T/P/G/U/X/C 1838; P/U

TRANSITIONAL SHIPS
Medea   1832; A/B LP/G/U 1847-8; L
Rhadamanthus   1831; A/B L/X/P/G/U 1836; P/G/U 1841; U
Salamander   1831; A/B LP/G/U 1836; S 1838; LP/G/U/X 1871; P/HX/G/U/R
Phoenix   1831; L/X/T A/B P/G/U
Hermes   1834; L/U/T/X Lengthened 1840; L 1842; P/G/U/S 1855; Cabins
Volcano   L/P/G/U/X 1845; P/U as floating factory 1854; P/U/F
Acheron
Megæra   L/P/G/U/X A/B P/G/U
Hydra   1837; L/X/P/G/U P/G/U/T port & stbd 1839; P/O/G/U A/B 1851; S
Hecate   P/G/U 1862; S 1863; GH
Hecla   P/G/U/T port & stbd
Alecto   1839; L/X/P/G/U/C 1860; G 1862; S 1863; GH
Ardent   G 1841; S 1850; S 1862; S
Polyphemus
Prometheus

GORGON AND CYCLOPS
Gorgon   1836; L/P/G/M/U/T/X A/B H/G/M/U 1839; P/H/G/M/U
Cyclops   1838; L/P/HG/M/U/X

FIRST CLASS SLOOPS
Stromboli   1838; L/X/P/H/G/U/S L/P/G/U 1858; S
Versuvius   L 1840; P/H/G/U
Cormorant   P/O/G/U
Driver   1840; L/P/HX/O/G/U/X P/HXO/G/U
Geyser   X/L 8in gun arrangement
Growler   L/X/P/G/U A/B stern
Eclair
Spiteful   8in gun Deck planking 1874; P/HXO/G/U
Styx   L A/B P/H/G/U
Vixen   L/X
Devastation
Thunderbolt   L/P/HX/O/G/U/X Paddle box boat. Tender
Virago   L/X/HX
Sphynx   P/HXO/G/U 1883; HX/O/G/U
Bulldog   1844; L/HXG/M/U P/G/M/U A/B P/HX/G/M/U
Fury   P/HX/M/U
Inflexible   A/B P/HXG/M/U
Scourge   P A/B P/HXG/M/U
Basilisk   1847; L/X A/B L/G/U/S 1864; S 1965; cabins/H

SECOND CLASS FRIGATES
Firebrand   P/HX/G/U 1858; M/U
Gladiator   P/G/U 1858; P/G/HX/U
Vulture   L/P/H/U A/B P/HX/G/M/U
Sampson   1843; L/G/M/X/U 1846; HX
Centaur   1844; L P/G/M/U/S A/B R 1855? R
Dragon   P/HXG/M/U A/B P/HX/G/M/U
Avenger   1844; L/P/HX/G/M/U/X/S A/B L/P/HX/G/M/U
Birkenhead   1843; L/P/HX/G/M/U A/B as troopship P/HX/G/M/U/R/S
Tiger   1847; L/P/HX/G/M/U/X/C/S A/B P/HX/G/MUS
Magicienne   1847; L/P/HX/G/M/U/C A/B P/HXG/M/U
Valorous   A/B P/HX/G/M/U Paddle box boats 1870-5 P/HX/G/M/U
Furious   1848; P/HX/GM/U 1862; S

LATER SLOOPS
Trident   1843; L/HX/G/U/X
Buzzard   1847; L/C/P/HX/G/U/S
Argus   1849; L/X/P/HX/G/U 1874; P/HX/G/U
Barracouta   1848; L/X/P/HXO/G/U A/B cabins 1872; P/HX/O/G/U
Janus   1842; P/H/G/U
Antelope   1845; L/S
Oberon   S 1876; X
Triton   L/P/O/G/U

FIRST CLASS FRIGATES
Penelope   As sail, 1838 L/X/S 1842; L/S 1846; S 1850; GO/U/Q 1855; HX

Retribution   1842; L/P/HX/M/U/X 1844; S A/B P/HX/G/M/U 1857;
             S 1850; HX/G/M/U/S section of topside 1862; S

Terrible   1842; L/C/X/P/G/M/U/T/S 1847; S 1849; L/X/P/G/M/U/S

Odin   1844; L/X/P/G/M/U/X/C A/B L/P/G/M/U

Sidon   1845; L/P/G/M/U A/B L/P/G/HX/G/M/U 1862; S

Leopard   1847; L/P/G/M/U A/B G 1852; II

NOTES

A/B	As built
L	Lines plan
P	Profile (Profile of inboard works). A longitudinal vertical section of the hull showing internal arrangements with some details of structure.
H	Hold
X	Sections
O	Orlop deck
G	Lower deck
M	Main deck
U	Upper deck
Q	Quarter deck and fo'c'sle (or spar deck)
D	All decks
R	Poop (Roundhouse)
F	Forecastle when separate from quarter deck
S	Sail plan. Usually outline only, without rigging; often at 1/16in to 1 ft.
T	Frame plan (Elevation)
HX	Hold and hold cross sections
WTC	Watertight compartments
C	Specification; includes scantlings and quality of materials to be used.

Documents

Public Record Office

ADM 12   Lists all correspondence relating to specific subjects. Is of great value in establishing the chronology.

ADM 95   Controller's Office Miscellanea. General information on ships and building, reports on sailing, defects etc.

Further Reading

There are few books which can be recommended on this subject.

D K Brown. *Before the Ironclad*, London (1990). More detail on the technologies involved.

J Fincham. *A History of Naval Architecture* (1851). Reprint, London, (1979). Good where he is describing his own work. It was a controversial time and his comments on others should be treated with care but not ignored.

C J Bartlett. *Great Britain and Sea Power*. Oxford, (1963). An excellent review of the political and operational aspects of the navy of the paddle era.

G L Overton. *Marine Engines*, Science Museum Handbook (also catalogue), London (1935). Fine description of the magnificent collection in the Science Museum, London.

E C Smith. *A Short History of Naval and Marine Engineering*. Cambridge (1937). Much of value on early marine engines. It is accurate but there are gaps.

Index